—— **2ND EDITION** ——

KEEPING BEES
AND MAKING HONEY

2ND EDITION

KEEPING BEES
AND MAKING HONEY

David and Charles

ALISON BENJAMIN AND
BRIAN McCALLUM

Contents

Introduction

People are always surprised when we tell them we keep bees in our small city back yard. Is the reason for their incredulity the fact that beekeeping is a hobby associated with large country gardens and pastoral scenes? Our urban friends tend to think of a beekeeper as a middle-aged man sporting a beard and brown sandals. While the stereotype does exist, beekeeping is fast becoming a more popular pastime for younger people of both genders in towns and cities. It may be impractical for most of us to live the rural idyll, and indeed many of us would not want to, yet keeping bees can bring a little of the country-side into any garden.

At a time when a generation brought up on computers has less exposure to the outdoors – a condition dubbed 'nature-deficit disorder' – bees can also bring us closer to the world of flowers and trees whose nectar is the source of their honey.

Bees are fascinating, industrious creatures. They live together in a complex community and are incalculably important to the cycle of life, pollinating a large percentage of the crops we eat. When people ask – as they invariably do – what made us decide to keep bees, one of the main reasons is that we wanted to help the environment by providing a home for nature's master pollinator.

The growth of interest in beekeeping means there are many suppliers selling hives and equipment that make the hobby safe, easy and, with ready assembled components and a variety of labour-saving devices, far less

time-consuming than it used to be. It is no longer the preserve of those with lots of time on their hands – even a high-flying city executive can keep bees nowadays. Like cats, bees do their own thing and require only occasional attention. The busiest time of the year is early summer, when you'll need to add extensions to the hive to give the queen bee enough room to lay her eggs and the workers sufficient space to store their honey. But even then, they will only take up a few hours of your week. In the winter, you need do nothing except occasionally check that the hive hasn't fallen over and the bees' honey stores haven't become seriously depleted – and preparing for the year ahead is also a good idea.

Most beginners find that it is worth joining a local beekeeping association, where they can pick up tips from experienced beekeepers keen to share their knowledge. But a word of warning: there is a saying that if you get three beekeepers together you will get four opinions on how best to manage your colony, and this is not far from the truth. The art of beekeeping is constantly evolving and beekeepers continue to invent their own ways of managing their colonies. Therefore there is no definitive guide on how to keep bees.

The goal of many beekeepers is to get as much honey as possible from their bees and they measure their success based on the quantity harvested at the end of the season. This, however, is not our goal, nor is it the purpose of this book. What we hope to do is

introduce you to the basic principles of modern beekeeping and enjoying the fruits of the bees' labour. As well as honey, bees produce wax, which for centuries has been highly prized in candles, polish and cosmetics. With a little practice you will be able to make your own candles, furniture polish and even hand and body lotions.

Beekeeping is endlessly fascinating, from the variety of honey the bees produce – no two harvests ever seem to taste the same, even from the same hive – to the different behaviour displayed by different colonies. You may find that before long what started as a passing interest becomes something of an obsession. And, as with anything else, the more deeply you delve into it the more complicated it becomes.

However, another saying among the beekeeping fraternity is that bees don't read books and there are many instances of bees behaving in ways that surprise even the most experienced apiarists. So don't be concerned if your bees do something unexpected. We can only assume that they know best – after all, they have been looking after themselves for millennia without any help from us.

Above all, this book aims to illustrate the huge enjoyment you will derive from having one of nature's most wondrous creatures residing in your garden. We really hope it will whet your appetite for beekeeping and encourage and inspire you to bring bees into your life, wherever you live.

Alison Benjamin and Brian McCallum

Understanding bees

The bee is one of nature's wonders. As well as being its most effective pollinator, honey bees produce delicious and healthy food that humans have been eating for centuries. While they can't be tamed, they can be given shelter and protection from the rain and cold. In return for providing them with a home, we can find out about the intricate workings of a colony, discover the joyful art of beekeeping and enjoy some of the bee's sweet-tasting food stores. In addition, a hive is an attractive feature in any garden or outside space.

History

Honey bees and humans

Humans have revered bees for centuries. Not only did they supply us with one of the first natural sweeteners, versatile medicines and precious commodities, but the workings of the hive and its residents' industrious behaviour have provided endless fascination as well as countless lessons. Political theorists have based numerous models of society on bee colonies, businessmen have drawn inspiration from their division of labour to develop management practices and the construction of their hexagonal honeycomb home has taught architects about the principles of design.

We know that beekeeping had begun by 30BC, because the Roman poet Virgil included a practical treatise on the art in his *Georgics*, in which he vividly describes the workings of a hive:

Aware that winter is coming, they use the summer days
For work, and put their winnings into a common pool.
Some are employed in getting food, and by fixed agreement
Work on the fields; some stay within their fenced abode,
With tear of daffodil and gummy resin of tree-bark
Laying their first foundation of the honeycomb, then hanging
The stickfast wax: others bring up the young bees, the hope
Of their people: others press
The pure honey and cram the cells with that crystal nectar.
Some allotted the duty of sentry-go at the gates....
Relieve incoming bees of their burden, or closing ranks
Shoo the drones – that work-shy gang – away from the bee-folds.
The work goes on like wildfire, the honey smells of thyme.

No wonder Virgil describes the hive as a 'tiny republic'. No wonder, also, that throughout history, from Aristotle and Plato to Erasmus and Marx, the bee colony has been held up in support of wildly different forms of human society, from absolute monarchy to republicanism and socialist utopia.

The bees' relentless toil has been proffered as an example for humans to emulate throughout the ages. The expressions 'hive of activity' and 'busy bee' have entered the English language, while the name of the non-working male, referred to as the 'lazy yawning drone' by Shakespeare in *Henry V*, is now used to describe a slothful individual. It is no coincidence that in the books of PG Wodehouse the Mayfair establishment where fashionable young men waste their days is called the Drones Club.

Right *A 15th-century portrait of Virgil (70–19BC) writing the 'Life of Bees' from* The Georgics.

The worker bee's industriousness was also embraced by the early colonizers of North America, who brought the honey bee with them from Europe. While peace has its olive branch, industry has its beehive, and US companies were quick to choose the bee and its home for their logos. The midwest state of Utah, whose motto is 'Industry', is known as the Beehive State.

Despite their heavy workload and their chastity – save for the queen and a few drones – bees have also been closely associated with love and sex. Golden, dripping honey has symbolized the delicious sweetness of both love and lust since biblical times: the Old Testament *Song of Songs* contains the lines:

Thy lips, O my bride as the honeycomb
Honey and milk are under thy tongue.

And the expression 'the birds and the bees' is still a euphemism for the mechanics of sexual reproduction.

For centuries, however, we got the gender of the big bee in the hive wrong. Just as it was assumed that human rulers had to be male, so the same prejudice was applied to the beehive: 'They have a king,' wrote Shakespeare in *Henry V*. It wasn't until English bee expert Charles Butler published his famous thesis, *The Feminine Monarchy*, in 1623, that the hive was conceived of as a matriarchal kingdom. Butler's ideas were confirmed a century later by Dutch scientist Jan Swammerdam in his *Book of Nature*.

For as long as humans have been studying bees we have also been utilizing the products of their labour. As many as 3,500 years ago the ancient Egyptians were making beeswax candles to provide artificial light. In the 12th century candlemaking became a profession practised by wax chandlers and there is still a Worshipful Company of Wax Chandlers in the City of London. Their main customer was the church; wherever Christianity spread in Europe, so did beekeeping and candlemaking. It is said that in Britain beekeeping never recovered after the Reformation which destroyed the monasteries, as monks and nuns were among the most enthusiastic apiarists. The bee then became associated with the Catholic church and from the 17th century was an official symbol of papacy. Not until 1900 were Catholic churches allowed to use non-beeswax candles.

Honey, beeswax and mead also had close ties with monarchs and landowners, who would accept them as payment of taxes and tithes. The practice stretches all the way from the ancient Aztec rulers of Mexico to feudal Welsh kings. The Greeks minted coins with bees on them and taxes were later demanded on bee products. In 16th-century France, beekeepers destroyed their hives rather than pay higher taxes.

In Greece and Egypt early hives were made of pottery and shaped like giant thimbles. The Romans used hives made from all manner of things – logs, bricks and even dung. In Europe, from the Middle Ages, wicker skeps were the predominant type. As they would rot if exposed to bad weather, special holes were made in walls in Britain and Ireland to shelter them. These 'bee boles' can still be seen in grand 17th- and 18th century country houses, such as Packwood House in Warwickshire. It was not until the invention of movable frame hives around the mid-1800s that modern beekeeping began.

Below *A 19th-century engraving entitled a 'study of different bees'.*

PLATE 10

HONEY BEE. 1.Worker. 2.Male. 3.Queen. 4. 5.COMMON HUMBLE BEE. LAPIDARY BEE. 6.Male. 7. Female. 8.MOSS or CARDER BEE.
9 DONOVAN'S HUMBLE BEE. 10.HARRIS' HUMBLE BEE. FALSE HUMBLE BEES. 11.Apathus Vestalis. 12.Apathus Rupestris.

Bee species

Entomologists, with their increasingly sophisticated technology, have identified more than 20,000 species of bee worldwide. The bees we are most likely to see in our fields and gardens represent just a handful of these. But what bees all have in common is that they eat food only supplied from the plant and in the process pollinate.

Solitary bees

The vast majority of bees are not into communal living. They have flying lives of no more than six to eight weeks and spend most the year as pupae in a cocoon. All the females of these species are fertile; once mated, a female lays around one male and eight female eggs, each in its own cell, in a small burrow or tunnel. The burrows can commonly be found in the ground, soft mortar, wood or the hollows of reeds and canes. The larvae spend most of the year developing in their individual cells; they emerge only as their food source is about to blossom. The first bee to emerge is the male, which quickly mates with any

females he may find. He then dies, leaving the female to look for a new nesting site to continue the life cycle.

Unlike the bumblebee and the honey bee, solitary bees usually feed on one plant, so when the plant's season ends the bee's food supply dries up and it dies. As a result, different species are seen at different times of the year.

You can encourage solitary bees to nest in your garden by installing a nesting site comprising a bundle of thin tubes. These can be made from 15cm (6in) lengths of 6mm (¼in) bore bamboo canes contained in a length of guttering. Attach this to a fence or wall in a sunny position around head height and with luck the bees will nest there. Some garden centres now sell ready-made bee boxes (below left) alongside their bird box selection.

Bumblebees

The bumblebee (*Bombus*), of which there are around 250 species worldwide, is usually larger and rounder than other bees. It has a thick furry coat, which allows it to fly in colder weather, so in temperate climates it is generally the first bee to be seen in spring. Bumblebees live in small colonies, with populations numbering from 50 to 400, and a queen who produces female worker bees.

Only the queen bumblebee survives the winter by hibernating in the undergrowth. In the spring she emerges and searches for a suitable nesting site in which to lay her eggs and build her colony. This nesting site will generally be below ground level in old mouse holes, burrows in the undergrowth or even cracks of concrete in the street. The queen quickly builds the wax cells to accommodate

her sterile, female offspring and as the population grows these worker bees gather the pollen and nectar for the colony.

Near the end of the summer, the colony raises virgin queens and males. They leave the nest to find a mate and while the males die after mating, the successful queens will hibernate soon afterwards. The vacated colony will die as winter approaches, then spring will see the cycle repeat itself with the next generation.

The buzz of bumblebees gathering nectar is a familiar summertime sound. But if the alarming decline in numbers is not halted it could become a thing of the past. In Britain alone, three species are now extinct and nine of the remaining 22 are endangered. Intensive farming methods have destroyed hedgerows and meadows (the bees' favoured nesting grounds) and increased use of insecticides has poisoned them.

You can make a patch of your garden more bumblebee-friendly by growing flowers that are rich in pollen and nectar and easily accessible to passing bees, and by having some plants that flower throughout the year. Providing a haven for the queen in winter by leaving areas of your garden untouched, especially with soft earth and coverings of leaves, can also help bumblebees to thrive. In spring, to encourage them to nest, put some dried moss or hamster bedding in an earthenware pot, then place the pot upside down in a hole in the ground. Cover the hole with a piece of slate or tile to stop the rain getting in but leave enough room for the queen bee to get under the tile. Hopefully she will find the pot and make it her home.

Honey bees

There are four main species of honey bee (*Apis*), which vary in size and characteristics, but all live in large colonies and store honey for winter supplies.

The species whose honey humans have harvested for centuries is the western or European honey bee, *Apis mellifera*, and it is this species that we will be concentrating on in the following chapters. So productive is this bee that it has been introduced all over the world to make honey and pollinate crops.

Honey bees are very well adapted to their role, with their long tongues (proboscis) designed to suck up nectar and their furry bodies to trap pollen. They are typically 15mm (just over ½in) long and live in colonies with populations measured in tens of thousands. They use wax, secreted from wax glands on the underside of their abdomen, to build the complex structures of connecting hexagonal cylinders we call honeycomb, in their nests. These nests are usually the size of a football but can be larger if space permits.

All bees are important to the planet because they pollinate much of the flora. Whether you keep honey bees in a hive or encourage their cousins to nest in your garden you will be helping to bring back an important insect that has been in decline.

Wasps

Many people group bees and wasps together in their minds because they look similar, buzz and can sting. But though distant cousins they are very different creatures. Wasps are omnivores and will kill for a meal, whereas bees eat only nectar and pollen and are quite passive. It is the wasp that hangs around your plate when you are having a barbecue, searching for any scraps of meat or fish, or diving into the jam pot while you're enjoying scones and cream in the sunshine. Wasps are persistent and a nuisance. Bees, on the other hand, are not interested in your food.

Wasps don't have wax glands, so make their nest from a mulch of wood fibres akin to papier mâché. They can sting more than once, whereas honey bees die after one sting, making them less likely to attack.

Though wasps get a bad press, and people often ask 'What are they for?', they do have a useful role to play in early summer, eating aphids, small flies and any decaying matter that is lying around in the garden.

PROPOLIS

Propolis is a resinous substance that bees collect from the barks of many trees and plants. It is sticky at 35°C (95°F) – hive temperature. It is usually brown or yellowish brown in colour. Poplar, birch, beech, horse chestnut and sunflowers are common sources.

The resin is used by the bees:
- as a glue to bind the nest together and seal up holes and gaps
- as an antiseptic when lining the cells to protect from moulds and infections
- to mummify dead intruders such as mice which are too big to be removed from the hive.

It is used by humans:
- to varnish the woodwork of hives
- as a health product with antiseptic properties, often sprinkled over breakfast cereal.

The great violin maker Stradivarius also favoured it to varnish his instruments.

The caste system

Within a colony of honey bees there are three castes: queen, worker and drone. Each has a distinct function to perform within the colony.

The worker bee

During the summer, when there is ample food around, a healthy colony of bees will typically be made up of one queen bee, a hundred or so male drone bees and 50,000 female worker bees. It is the workers that we see dancing around the flora of our gardens.

All matters relating to the wellbeing of the colony, except for egg-laying, are handled by the worker bees. Their mission in life is the survival of the colony. In pursuit of this goal, they work together through a complex communication system. Their duties around the hive change with age (see 'The Life Cycle') and they have wax glands to produce the wax that is made into the honeycomb.

Workers communicate to each other by touch, smell and vibrations. The queen bee emits a pheromone (scent) that binds the colony together. Worker bees will often be seen standing still at the hive entrance, with their bottom in the air, beating their wings rapidly, wafting their scent into the air to guide their fellow workers back to the hive.

The drone bee

The drones are the males of the colony. They are physically larger than the workers, squarer in shape, with big eyes. Their sting has been replaced with the male sexual organ, and they don't have wax glands. Their main role is to search out, find and mate with a queen.

The colony will produce a few hundred drones in the spring in anticipation of the needs of a virgin queen. Interestingly, the drones are hatched from unfertilized eggs, which means they don't have a father. Since they are larger than the workers, their cells are larger and can be seen in the hive quite easily. When they are mature they will take to the sky to find a receptive queen with whom to mate. The mating process actually kills the drone, but by that time he has ensured that his genes are passed down to future generations.

Above *A drone bee takes to the sky to find and mate with a queen.*

The queen bee

We give the queen bee a royal title but in fact she has no real powers over the colony. She doesn't rule or command them, but without her the colony would die. Her role is to lay eggs and she does this continuously in the dark for the duration of her life, reaching 2,000 a day at her peak. She even has a group of attentive maids who feed, clean and clear up after her so that she doesn't need to disrupt her egg-laying with such trivial matters.

The queen emits a variety of pheromones which bind the colony together. The exact purpose of all these scents isn't clear, but we know that one of them reassures the colony.

Right *The queen bee surrounded in the hive by an attentive team of workers who feed, clean and clear up after her.*

The anatomy of a honey bee

The Nasonov gland emits another pheromone that says 'Here we are' and is used when the bees are swarming or to guide a forager home.

The wax squeezes out as a liquid from the wax glands on the underside of the abdomen, where it solidifies into tiny scales of wax.

The hypopharyngeal and mandibular glands, inside the head, are used to make royal jelly (see 'The birth cycle').

The sting and the associated poison are used in defence. The sting is barbed, so that it can attach itself to the fibres of the recipient. When the bee pulls away, her insides stay attached to the lodged sting, subsequently tearing her in two. When a bee stings, the sting emits a pheromone which alerts her sisters to the danger and directs them to sting in the same location.

A honey bee has five eyes: three simple eyes at the top of the head, which she uses for orientation, and two complex eyes, which are for seeing. A bee sees colour and shapes: it doesn't see slow-moving objects well. Fast, jerky movements will attract much more attention – something to keep in mind when working around bees.

Abdomen

Thorax

Head

The antennae are used for smelling and are very sensitive to variations in odour.

The two hind legs hold the collected pollen on the corbicula (pollen basket).

The mouth has two mandibles which act like teeth and work like pincers. The bee uses them to manipulate wax, to collect propolis and to bite.

The four wings and six legs are attached to the thorax – the powerhouse of the bee. The smaller hind wings are hinged to the more powerful front wings, allowing the hind wings to fold on to the body. This gives the bee more room to move around the crowded colony and get into small flowers for nectar. The wings can also be rubbed against each other to make a sound that is used in communication.

The honey stomach holds the nectar after the bees have collected it from the flowers. Enzymes act on the nectar, which starts the process of honey making.

The two front legs have antenna cleaners.

The proboscis is the tongue, which acts like a straw to suck the nectar into the honey sac. It is up to 7mm (¼in) long, good for getting deep into the flowers.

The two middle legs clean the wings, brush the pollen on to the corbicula and take the wax from the wax pockets on the underside of the abdomen.

The birth cycle

The queen bee, the mother of all the other bees in the hive, continually searches for clean cells in which to lay her eggs. She backs her way into one of the cells and attaches an egg to the back wall with a type of glue. She then moves to the next available cell to lay her next egg. She may do this up to 2,000 times a day.

Each egg, reminiscent of a small grain of rice, hangs horizontally from the cell wall. It takes three days for a larva to hatch from this egg and when it does, like all newborns, it is hungry and needs food in order to grow. Worker bees are on hand to feed the larva with the nutrient-rich substance we call royal jelly. This jelly – secreted by the workers as part of their saliva – is fed to all larvae for the first three days, after which those that have not been chosen to become queens are switched to a mixture of pollen and honey.

The larva now goes through a series of instars (moulting of the outer skin). With each stage it grows bigger and has to curl up into a 'C' shape to fit into its cell. After six days the pearly white larva straightens itself out and stops eating. The attending bees notice this change of behaviour and build a cover over the cell, using a mixture of wax and propolis to enclose the larva for the next stage.

During the next 12 days the covered larva pupates, spinning a cocoon and changing from a grub into an insect. When this metamorphosis is complete it eats its way through the cell cap, emerging as a young honey bee.

For a worker this process takes 21 days, for a drone 24. The queen, in her specially enlarged cell, develops more quickly and emerges from her cocoon after only 16 days.

Beekeepers soon learn to differentiate between each of the brood stages. Easiest to see is the capped pupa cell, but with experience you will learn to recognize the fat white larvae and the tiny eggs. This skill will become important when you need to check the health of the colony.

The area within the hive that contains the eggs, larvae and pupae is called the brood nest and is oval in shape. This area increases as the queen lays more eggs in the warmer months but the overall shape stays the same.

Left The birth cycle of the honey bee.

ROYAL JELLY

Royal jelly is a white, jelly-like, protein-rich substance made by the bees from the hypopharyngeal glands located in their heads. In addition to using it to nourish the young larvae, attendant worker bees also feed royal jelly to an adult-laying queen.

Royal jelly is thought to have rejuvenating powers for humans and is a much sought-after ingredient in cosmetics. But realistically, as an amateur beekeeper with a few hives, you are unlikely to be able to harvest it.

The life cycle

Workers

The duties that a worker bee carries out throughout her life change with age and physical maturity. A newly emerged bee is not fully developed – her sting and wax glands, for example, have yet to mature.

1. She begins her working life as **a cleaner** straight after birth. She will inspect, clean and polish cells for the queen to lay in.

2. A few days later her task is to **feed the older larvae** with pollen and nectar. When her own hypopharyngeal and mandibular glands have developed, she can start to feed the younger larvae with royal jelly.

3. The wax glands develop next, so she can begin to **cap the larvae** and honey cells, and use the wax to repair and build new comb.

4. Two weeks old, and she can **collect nectar and pollen from returning foraging bees** and place this food in the store cells.

5. 18 days old – three days before she begins to fly outside – she will start to **guard the entrance to the hive**. Her sting has fully developed by now and her mandibles are strong. She spends time at the entrance smelling the returning bees to make sure they are from her colony. If they are not residents, like a bouncer at a nightclub she bars their entry. If the chancer won't take no for an answer, she will resort to using her sting to kill the intruder, but will die in the process.

6. During these first three weeks in the hive, the worker's normal tasks may be interrupted by more urgent ones, such as **ventilation**: flapping the wings rapidly circulates the air and helps to regulate the temperature of the hive. A bee also fans nectar to evaporate the water content, turning the nectar into honey.

7. Three weeks old – she is ready for **foraging**. She will take a few orientating flights to get her bearings and then spend the rest of her life – a further three weeks – collecting nectar, pollen, water and propolis.

This is draining work and is fraught with dangers. She may fall prey to spiders and birds, her wings get battered from hitting the flora and she will travel miles carrying a load of pollen and nectar weighing as much as she does. In theory a bee can live for several years, but in practice, six weeks after being born, most workers have died of exhaustion.

However, a worker who is lucky enough to be born in late summer could be one of the 10,000 bees who live through the winter, huddled together eating the stored honey. Her luck will come to an end when the foraging season returns in the spring and she is among the first out to search for new food.

Drones

The role of the drones, the male of the species, is to mate with a queen. When they are not out flying looking for a queen on her mating flight, they spend their time lazing around, eating honey and making a mess of the hive. They rely on the colony for their food and shelter, and though they may help with ventilation they do little else around the home. They can live like this for most of the summer and, unlike their sisters, they can migrate between colonies while they look for their prize. They mate in flight, but this is the last action of their lives. As the drone dismounts from the queen his sexual organs are ripped from his body and he falls, dying, to the earth. Any drone who fails to mate and is still in the hive towards the end of the summer has outlived his welcome. The colony can ill afford to waste its stores of honey on non-productive members, so these surviving drones will be evicted by their sisters and, once outside the hive, will soon starve to death.

The queen

Unlike other bees, a queen can live for years. She spends practically all of that time in the darkness of the hive, venturing out on only two occasions: once at the beginning of her life to mate and later to swarm.

Any fertilized egg can become a queen if the colony decides it needs to raise a new one. The chosen egg's cell is enlarged and built to hang vertically (worker cells lie horizontally). The workers feed the growing larva intensively on royal jelly (every five minutes for 5 ½ days), which stimulates it to develop into a queen. The metamorphosis takes 16 days and, after a few days of acclimatizing, the new virgin queen will leave the hive to mate.

On this nuptial flight the queen's phero-mones attract the drone bees and she mates with up to 12 of them until her spermatheca (the sac that holds the sperm) is full. On returning to the hive she begins her life as an egg-layer. When she eventually starts to lose her egg-laying abilities due to old age or illness her pheromones change and this signals the workers to raise a new queen. Sixteen days later the new queen will kill her mother and take over the hive, a process called supersedure.

Superorganism

The occupants of a honey bee colony live together in a tight social network, dependent on each other and unable to survive if they are separated from the social group. They work together in harmony for the good of the whole colony, sometimes even sacrificing themselves to ensure the whole honey bee colony survives. This type of social structure is quite unusual in the animal kingdom and is called a 'superorganism'. This means that a colony of bees, not the individual bee, is the complete animal. So, the adult bees we see flying to the flowers are just one part of a honey bee colony. The other parts are the queen bee, the male drones, the wax honeycomb on which the colony lives, the brood which is the developing bees, and the food stores – the pollen and nectar – which are the colony's energy source and needed at all times. These elements make up the whole being.

It is easier to understand the concept of a 'superorganism' if you compare the colony of honey bees to a human body; the cells and organs of our body are the worker bees, the wax comb is the skeleton and skin and the queen and the drones are the reproductive organs.

The skill and art of beekeeping is keeping our honey bee colonies healthy and allowing them the opportunity to flourish and live in balance with nature. It takes time and experience to learn, but knowing how the colony itself works gives you a head start. If, for example, a colony's energy stores are depleted because the adult worker bees can't fly due to inclement weather, we may need to give our colony supplementary food. We want our colony to be as stress-free as possible, giving it the best chance of survival and growth.

Communication

Our honey bee colony resides in a dark cavity. As beekeepers we give the colony a sheltered, dry and dark place to live: the hive. The main way the colony communicates in the dark is through specific odours (pheromones). The worker bees' sense of smell is very well developed and there are numerous pheromones within the colony which trigger a different behaviour. The larvae for example emit a smell which says 'I'm hungry', triggering the adult bees to feed the larvae with pollen and nectar.

The colony also uses vibrations to tell each other what is happening. Adult bees will hold on to the wax comb and shake their bodies. The comb transmits through the wax structure this vibration which can be picked up by other adult bees even when they are far away from the shaking bees. This type of communication is used for many purposes, for example, a bee that wants to be preened will stand and shake, which alerts neighbouring bees to come and preen her. This shaking behaviour can be seen throughout the colony nest. The waggle dance which communicates the location of a source of food is a particularly sophisticated example of this bee language, described in 'How bees make honey'.

We are still discovering how the colony communicates but we know that each communication triggers a reaction. This helps

Right *The worker bees with their queen who emits a phero-mone that binds the colony together.*

the colony deal with events that happen around it. If many bees are saying there is lots of available food on the flowers then this triggers more bees to go out and forage. As the foragers return to the hive they in turn recruit even more bees to forage and the colony quickly exploits the available food source with its increased workforce. When the source of food dries up fewer foragers are needed so they turn their labours to some other task within the hive, for instance ventilating the hive.

Since the honey bee brood needs to be kept at a constant temperature, the colony

> **We are still discovering how the colony communicates but we know that each communication triggers a reaction.**

regulates its environment by warming it up if it is cold outside and cooling it down if it is too hot. The worker bees achieve this by fanning their wings to circulate and expel the hot air or by shivering which creates heat to raise the temperature of the nest. It is this communication and triggering action that keeps a colony in balance and in harmony with the natural world around it.

In balance with nature

As the seasons change, the colony changes too. It is tuned to the sun – the forager bees need the sun to navigate – so the colony recognises that winter is coming by the shorter daylight hours and as a result it begins to make changes that will allow it to survive the winter months when it is too cold to forage and there is a finite amount of food available in the hive. These changes include the queen reducing her egg-laying activities so there are not nearly as many hungry mouths to feed. The colony's worker

bee population reduces in size until it reaches the optimum number that can survive by feeding on its winter stores of honey. This is usually around 10,000 worker bees.

It is the law of nature that a species will seek to reproduce. Because a honey bee colony is the organism, it is the colony that will reproduce through creating another honey bee colony by a process of splitting itself in two. This process is known as swarming and it generally occurs in the spring and early summer, a time when there should be an abundance of available food which allows the new colony the opportunity to survive and grow fast. We talk more about this in 'The second year'.

By becoming a beekeeper we begin to understand the amazing interactions that take place within a honey bee colony and how the colony reacts to the outside environment. We realise that a colony is only a healthy animal when all the parts are working together as a whole unit. It is not just the queen, workers and drones but the comb, brood and food stores that together make up this extraordinary animal: the honey bee colony. It makes its own home, regulates its environment, communicates required changes and coordinates its life with the natural world around it. And in the process of living and collecting its food (which supplies us with wonderful honey) it pollinates the plants that feed so many living creatures – including us.

What to consider

Before deciding to keep bees you need first to be clear about your aims. Ask yourself whether you are interested in profit or whether you want to keep them purely for the pleasure of doing so. You should also weigh up how much time and space you have at your disposal. Little of either need not preclude you from keeping bees, but it will mean that owning more than one or two hives is out of the question and it is unlikely you will produce enough honey to make your hobby pay for itself.

Space

Keeping bees is traditionally thought of as a rural hobby, the preserve of people with large country gardens or acres of land. But it is becoming more popular among urban residents and suburb-dwellers. One reason for this is that a beehive doesn't actually take up much room – it can fit on to a standard paving stone – so you don't need a large garden to accommodate it. Hives can be kept in the smallest of back yards, on roof gardens and even on balconies (right). More important than a large space in which to put the hive is adequate storage for all the paraphernalia that goes with beekeeping (see 'The kit'). This might be a small outside shed, a large cupboard, or space in the basement.

The fact is that urban bees do very well, possibly because there is more biodiversity in towns than in the countryside these days. It is said that, because of our increasingly industrialized agricultural practices, 0.2 hectares (½ acre) of overgrown churchyard is home to a greater variety of plants and wildlife than 16 hectares (40 acres) of monoculture arable land. In addition, city gardens and their flora are as diverse as their owners and produce an abundant variety of year-round forage. Add to this mixture the parks, tree-lined roads and undisturbed railway sidings and you will find that bees are more than happy with an urban environment.

Siting the hive

The location of your hive is the first thing to think about. In this respect, smaller gardens present some issues that larger properties generally don't, such as the hive's proximity to your and your neighbour's back door, but the principle for finding the best place to put your hive is the same for everyone. You want comfort for yourselves and the bees, so choose a spot that is dry, sheltered, undisturbed and far enough away from people that neither you nor the bees are interrupted in your daily business.

When they leave the hive to forage, bees naturally fly upwards in a series of increasingly large spirals. This orientates them and imprints the mental aerial map they need for their return journey. When they return they are heavy with their booty and will fly directly back to the hive. If you are putting a hive close to where people walk you can encourage the bees to fly above head height and down to the hive by placing some 1.8m (6ft) hedging or fencing around the hive. The bees will soon learn that it is easier to fly directly down to the hive rather than approaching at a low level and then having to climb to negotiate a high barrier with a heavy load.

Bees like a bit of morning sunshine, so if possible site the hive in an area that catches the early rays. This warms the hive and lets the bees know that the day has begun: getting them out of the hive early will increase their foraging time, which should result in more honey at the end of the season.

Like most creatures, bees like to be sheltered from the elements. The hive itself will protect them from most of what nature throws at them, but don't put it in a windy spot and don't have the entrance facing the prevailing wind. Bees are strong and can fly in windy conditions, but no one likes a draught.

Bees like a bit of morning sunshine, so if possible site the hive in an area that catches the early rays.

Cold

Although bees regulate the temperature inside the hive, try to keep it somewhere that gets neither too hot nor too cold. If your chosen site is in the sun all day, give the hive some shade either by planting some foliage or by putting up a fence. Avoid sunken areas, such as a hollow or dip, where cold air can get trapped – this will mean the colony has to work harder at keeping the brood at the required 35°C (95°F).

A hedge or some form of wind break is desirable if the hive is to be in an open spot. If the area is prone to damp, place the hive on a high stand (see 'Hive components'). Bees don't mind the cold, but they dislike the damp. Although shading from trees is helpful during the summer to prevent the hive overheating, the branches mustn't droop so close to the hive that they will bang against it when the wind blows. This will disturb the bees, and in a storm branches could even fall on the hive and destroy it.

Water

If the area you live in doesn't have a natural source, you will need to provide some fresh water. This can be from a water feature in your garden or a pot filled with sodden peat. Once the bees have found the supply they will be frequent visitors, but as they can't swim you need to make sure they have something to stand on while they drink – otherwise they are likely to drown. Floating sticks will suffice. Keep the supply of water a few metres (yards) from the hive so the bees can defecate without contaminating the water supply. They are very clean creatures and only relieve themselves outside the hive, even in winter when they have to wait for a warm day before they can venture out.

A bee shed

You can even keep bees in a garden shed. Bee houses, as they are known, are especially common in northern Europe. They are practical, as the equipment, clothing and sundries can all be kept inside the shed too. Just make a slit in the side wall of the shed and place the hive on a shelf with the entrance facing the slit. On the outside, attach a piece of wood – 30cm x 9cm x 2.5cm (12in x 3in x 1in) is fine – as a landing board to aid the bees' return. Depending on the size of the shed, a number of hives can be kept this way side by side – if you have more than one, paint

Below Left If the hive is in a windy location, you may need to put a weight on the roof to keep it in place.

Below A bee shed is a good way to keep a number of hives. If you paint the landing boards different colours it helps the bees to find their way home.

each of the landing boards a different colour to help the bees return to the correct hive. The downside of this approach is that you forgo the aesthetic pleasure of having a hive to look at in your garden.

Two hives

It is quite common for the novice beekeeper to become so enthralled with the hobby that they soon think about increasing their apiary, so it is worth considering a site that may have the space to accommodate more than one hive. All the same rules apply, but you should angle the second hive's entrance to face in a slightly different direction, so that the bees know which hive to return to, and place it at least 1m (3ft) away from the first.

Out apiary

If you don't have anywhere suitable for keeping bees, consider asking a local land-owner if you can keep a hive on their land in return for some honey. Arable or animal farms, part of a cemetery that is not open to the public, the grounds of a retirement home, public allotments, secluded areas within local parks, or brown field sites to which the public does not have access, such as overgrown land around disused warehouses or factories, could all make suitable locations for a hive or two.

In addition to supplying jars from the honey harvest, you could also offer to give a talk about beekeeping to residents of the retirement home, the church congregation or users of the park.

If you plan to keep a hive in a location far from your house, there are additional factors to consider. You want it to be easily accessible, as there will be times when you have to take extra equipment to it and other times when the honey is ready for collecting that you will have to carry a fairly heavy load back. So don't put the hive anywhere that forces you to walk a long way, or where there are narrow gates and openings to negotiate when your hands are full.

Keep the hive away from public footpaths and thoroughfares. As well as the risk of people being stung, hives can sometimes be the target of vandalism, abuse and theft.

tip
If the area you live in doesn't have a natural source of water, you will need to provide fresh water.

Hives at the workplace

Beekeeping is becoming a popular pastime for office workers in their lunch hour. The honey bees' industriousness is legendary and it has been embraced as a symbol of hard graft, thrift and teamwork through the ages, from Roman poet Virgil to the US state of Utah and companies including IBM. Now some workplaces are embracing rooftop beekeeping to encourage their employees to learn about the bees.

Considerations for workplace hives are similar to other locations: do you have the space, is your preferred location away from people, is it windy, is their enough room for two hives and are the hives in a secure spot? But there are some additional considerations: do you need permission from the freeholder of the property from whom your office is rented to keep bees, and do you need to produce a risk assessment report for the insurance company?

If you are proposing to site the hives on a workplace roof, the freeholder, or the company that manages the office for the freeholder, may be nervous that staff such as window cleaners or contractors who may need to get access to the roof, could be stung and take legal action. Often such fears are the result of ignorance about bees' behaviour and they can be allayed by educating the relevant people. You may need to organise a talk from an experienced beekeeper to outline the possible risks involved.

You may also have to produce a written risk assessment which will list all of the potential risks such as getting stung and how you prepare to mitigate the risks by, for example,

ensuring staff, contractors and any visitors to the hive wear bee suits.

Even if your employer owns the building in which you work, they will still have to inform their insurance company or broker about a potential hive. The insurer could ask your employer to produce a disclaimer for all staff who volunteer to become beekeepers. Staff would need to sign the disclaimer to indicate their understanding and acceptance of the potential risks, such as getting stung or falling off the roof, and to agree to a number of conditions, from attending training sessions to wearing appropriate beekeeping gear and not going beyond the permitted boundaries on the roof terrace.

Insurers could insist that barriers are erected along the roof at the company's expense and for the company to inform contractors where the hive is located and to offer them bee suits when they need to access the roof.

Many companies do not have access to the roof of their building, or it may not be suitable for a beehive because it is 20 storey's high or in a wind funnel. If this is the case, your company could think about sponsoring a beehive in a community-run apiary nearby and offer their employees the opportunity of beekeeping in their lunch hour. In fact, sponsoring a community beehive instead of having your own could fit into a company's corporate social responsibility (CSR) strategy; employees would be volunteering in the local community by helping at the community apiary as well as learning about bees and the environment.

Some companies have taken this CSR a step further by supporting employees to volunteer with local young people, or homeless people who learn about beekeeping and in the process develop new skills, confront their fears and increase their confidence.

If your workplace is in a city centre, or financial district, surrounded by offices and what appears to be little green space for miles around then bee forage may also be an issue. Keeping a hive should go hand in hand with improving the green infrastructure to making it more bee-friendly. Although bees may fly for three miles for food, a closer source of year-round pollen and nectar from bee-friendly trees and shrubs will improve the health of your bees.

Your company may be part of a business improvement district (Bid) which aims to improve the area in many different ways, including expanding the green infrastructure by helping companies to introduce green roofs or green landscaping.

Wherever a company decides to locate its beehive, the main challenge of keeping bees at work is actually employees' time commitments. However enthusiastic you and your colleagues may be, you may find you are just too busy with meetings, business travel and other work to fulfil your voluntary obligations as a beekeeper. For this reason, it is useful to have a team of trained beekeepers so that they can fill in for each other. Storage space for equipment and suits will need to be found and the team will be responsible for keeping equipment and suits clean and ordering new equipment when necessary.

School apiary

For schools thinking of keeping a hive to teach their pupils about the biology of the bee or their role in the environment, considerations in addition to those already mentioned include reassuring concerned parents that the risks of their child being stung are minimal. Risks of stings can be mitigated if the hive is sited away from playing fields or playgrounds, only accessible through a locked gate, enclosed by a seven-foot high mesh-netted fence, and pupils have to be accompanied by a trained member of staff and wear a bee suit. Staff should be first-aid trained and parents of pupils wishing to participate in beekeeping should sign a disclaimer that to their knowledge their child does not get a severe reaction to bee stings (see 'Routine').

Ideally, a group of teachers should be trained beekeepers responsible for maintenance of the school apiary and for beekeeping lessons. They need to share the workload and the responsibility. A bee-sitting rota should be established so teachers can take it in turns to inspect the bee hive during any school holidays in the spring or summer.

Many schools now keep bees. They often have close links to their local beekeeping association, which may be able to provide a mentor to assist the teachers who are looking after the hives. Some of the members of the association may have children at the school and could form a parents' beekeeping group to provide additional support. Some schools host hives on their land. These are paid for and maintained by a local beekeeper who will also give talks to the pupils.

Schools with hives have integrated the bees into lessons. But some have witnessed the benefits of keeping bees far beyond the curriculum. By introducing their most disruptive and unruly pupils to beekeeping they have seen a significant improvement in the pupils' behaviour and concentration; the pupils must be calm around the bees and have finally found something they enjoy at school and are good at.

Below *Beekeeper at a school apiary in the City of London.*

Children

There seems to be something captivating about a hive. The foraging workers returning and telling the gang their news, or the queen steadily walking around the frames looking for a clean chamber in which to lay her eggs, can keep most of us intrigued, whatever our age. In addition, many children have been brought up on stories about a certain bear and his love of honey.

Bees are not, of course, pets in the traditional sense. You can't take them for walks, stroke them or cuddle them, but then again you don't have the chore of cleaning out their home as you do with a rabbit or guinea pig, and you do get that sweet honey as a treat.

Bees can become part of the family and in today's society, they are a perfect antidote to 'nature-deficit disorder'. What better way for children and their parents to learn about the workings of nature together?

The worry of being stung is always present with bees, especially when there are children around. But some simple precautions will keep everyone safe and happy. Many bee suits come in children's sizes, so they can easily join in with the hive inspections. Children are often eager to ask questions and learn what's going on, and there is a good chance they will spot the queen on a frame before you do.

> The worry of being stung is always present, especially when there are children around. But some simple precautions will keep everyone safe and happy.

Honey-harvesting time is particularly fun for children: sticky, sweet and messy, just the sort of thing they love. They'll especially enjoy spinning the extractor to get the honey off the frames (see 'How to harvest honey'). Many schools have harvest festivals around this time, so children can also have the thrill of contributing a jar of home-grown honey.

And it's not just about honey. Candlemaking is a craft that you and your children could learn and enjoy together (see 'Beeswax candles').

Pets

Many people cite their pets as a reason why they can't keep bees. You may be worried about cats or dogs getting stung, but cats at least are more sensible than you might give them credit for. Although the roof of an empty hive – before the bees arrive – may be a favourite sun trap, they will soon learn to steer well clear once the bees move in. Dogs can be a bit more inquisitive, so be sure to shut them in the house when the hive is open for inspection.

Horses can kick hives over, accidentally or on purpose, so try not to have bees in the same paddock. If this is unavoidable, protect the hive with some sort of barrier.

As you will find out, bees are generally passive creatures who are content getting on with their own lives and happy to coexist with you and your pets.

Below *Children are fascinated by bees, and will enjoy taking part in hive inspections and maintenance. Bee suits come in children's sizes, so beekeeping can be a safe, family experience.*

Routine

Beekeeping isn't a labour-intensive activity. For the most part bees look after themselves and our role is to make sure the colony is kept healthy and has the space it requires. After all, bees have been getting by without us for millions of years so you will find that, with little interference, they will happily produce all they need for their own purposes – and a bit of extra honey for you.

The work involved varies with the seasons. In winter there is little need to check on the hive. It will only require the occasional visit just to ensure that it has not been knocked over or a tree branch hasn't fallen on it. There will be some work to be done away from the hive such as making the frames and supers (see 'Hive components'), but this doesn't take long and gets quicker with practice.

During the warmer months the workload increases and at certain times weekly visits are called for, but even so one colony shouldn't take up much more than a couple of hours per week in the busiest period and an hour per month during the winter.

It is useful to keep a notebook and enter everything you have done or seen at the hive. Especially if you have more than one hive, it is surprisingly easy to forget what you did when and at each one. A notebook also helps you plan ahead and schedule the next visit.

Going away

One advantage of bees requiring only the occasional visit is that you don't have to rely on your neighbours' goodwill while you are on holiday. Even if you have to go away at the busiest period, in early summer, don't worry. With some planning and preparation, such as adding a super or two in case the bees run out of space (see 'Adding supers'), you can safely leave them to get on with things.

Cost

Small-scale beekeeping doesn't need to be expensive. The initial nucleus of 10,000 bees, a hive and all its components, bee suits, smoker

Above *It is useful to have a basket or container for all of your essential beekeeping tools, including a smoker, fuel and gloves.*

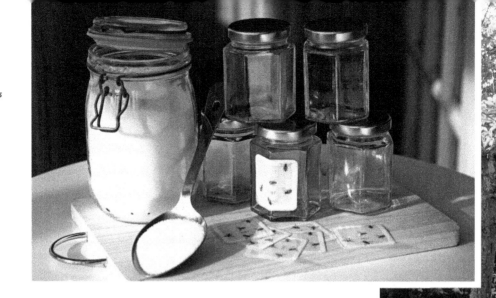

and other equipment are the biggest outlay. Thereafter, the running costs are fairly low: a few pounds of sugar, pots, jars and labels are the only likely extras. There are many suppliers selling a whole range of gadgets and mechanized apparatus who will gladly take your money, but it is not necessary to purchase absolutely everything. More pricey items such as honey extractors can be shared or hired from local bee associations.

Shop around for your first hive and equipment. Prices vary enormously, but as with most things quality merchandise lasts longer, so buy the best you can afford. Specifically, go for a cedarwood hive if you can (see 'Different types of hives'). There is a second-hand market, but make sure you are buying from a responsible outlet, as disease can be transmitted through used equipment. Make sure everything has been cleaned, scorched and sterilized. Your local bee association will recommend responsible resellers in your area.

Hives are really just modified four-sided boxes. Hives and frames can be bought either made up or as a flat pack. Flat pack is cheaper and is delivered more quickly (which can

prove crucial) if you're buying mail order, but it takes time to assemble and if you do not know one end of a hammer from the other it can be daunting at first. If you like carpentry, you can construct a hive from scratch using drawings and plans readily available on the internet.

Neighbours

Not everyone will be as happy as you are with the idea of a beehive in their vicinity. It is only courteous to alert your neighbours to the new arrivals who will shortly be moving in. Although there may be hostility initially, it should take only a few weeks for everyone to realize that bees keep themselves to themselves and are not interested in humans.

Try to allay your neighbours' fears and reservations by highlighting the benefits of keeping bees. Inform them that: bees are the best pollinators around, so trees, flowers and fruit in the neighbourhood will flourish; they give us gorgeous honey – and you will gladly give the neighbours a jar or two; and they are a wonderful way to see nature at work. If anyone in the area suffers from hay fever, impress on them the medicinal benefits of eating locally produced honey.

The major fear will be about being stung. Reassure them that bees are not aggressive, and because they eat only nectar and pollen, they are not interested in our food – unlike their carnivorous cousins the wasps. Bees are also quiet, don't smell and, despite some popular misconceptions don't chase after people in large, killer swarms.

BEE STINGS

Bees are not angry, nasty or hostile creatures and they are not sting-happy, because the action of stinging kills them. But they are defensive, recognizing when they are under threat and responding accordingly. As we are sometimes intruding into their home it is inevitable that they will try to sting. Most people react to a sting with a localized swelling followed by a couple of days of itchiness. The swelling can be large and red but this is completely normal. Some people, though, have a serious allergic reaction to bee venom (in the same way that some people are allergic to nuts) which can result in a life-threatening anaphylactic shock. If you are at all concerned about the reaction that you or your neighbours may have, seek professional medical advice.

Local bylaws

Check with your local council in case there are any ancient laws that forbid beekeeping in your area. This is extremely rare but not unknown.

Above It is a good idea to tell your neighbours that you will be keeping bees once you have decided where you are going to locate your hive.

tip

Try to allay your neighbours' fears and reservations by highlighting the benefits of keeping bees.

Housing your bees

How your bees are housed depends on considerations such as whether you want an aesthetically pleasing hive or one that is easy to use, and how much money you want to spend. The initial cost of a good hive and its components will be your major outlay but they will last a long time if they are looked after well. There is a wide range of hives on the market but most follow the single-walled box model. During the summer when the flowers are in bloom you will increase the height of the hive to give the bees additional space to store their honey. It is not uncommon for a hive to stand 1.5m (5ft) or more tall by the end of the season.

How a hive works

As we have seen, bees are social insects living together in a colony with a queen, thousands of workers and a few drones. The colony itself can be thought of as a living organism and each individual bee as a component part whose job is to keep the organism alive.

In the wild, bees live happily in dark, dry cavities with small defendable entrances. Inside this cavity they build their nest from self-generated wax. Workers shape the wax into connecting hexagonal cells that protrude from either side of a vertical sheet hanging from the roof. As a colony grows, the workers hang more vertical sheets – a large colony may have eight sheets, with the central ones being the longest and widest. The cells on these sheets are used for birth chambers, honey and pollen stores and resting rooms.

The distance between each sheet is constant and uniform and is about 9mm (⅓in) for the western honey bee. This distance is just enough for the bees to crawl around in and pass each other without getting crushed, but not so far apart that they lose the important bodily contact. This distance is important for beekeepers and is known as 'bee space'. More on this later.

Prior to the invention of the modern-day hive, when colonies were kept in upturned baskets called skeps, a colony was destroyed when the honey was collected. This wasn't good for the bees or the beekeeper.

Thanks to the work of some eminent entomologists in the mid-19th century, a better system was developed. This exploited the principle of the bees' natural behaviour of building sheets of comb. A number of frames were housed in a box and the bees used these frames to build their vertical sheets of wax comb. When a frame was filled they moved on to the next one. More frames could be added when needed. The advantage of this system was that the frame could be lifted out of the box without the colony being destroyed. This was called the 'movable frame hive' and although it was a combination of many people's ideas it has been attributed to the Reverend LL Langstroth, who patented it. His eponymous hive is widely used today, most notably in the USA.

Reputedly, the first box that was used to hold these removable frames was an empty champagne crate, and today's hives haven't changed that much. What you get when you buy a hive is a number of roofless and floorless boxes that sit on top of each other, each capable of holding ten or so frames. The boxes come in two depths, deep and shallow, usually known as brood boxes and supers. Brood (deep) boxes are so called because this is generally where the brood (developing larvae) is housed while the supers – Latin for 'on top' – are above the brood nest and are where the bees store their honey. The reason they are more shallow is so that they are not too heavy for us to lift when they are full of honey. The bees don't mind the size of the boxes or the frames.

A hive also has a roof, a floor and a stand. The tower they make has a footprint of around 60cm (2ft) square and a height ranging from 60cm–1.8m (2–6ft) depending on the season – you need to add extra floorless super boxes to save the incoming nectar as the summer progresses.

The art of beekeeping lies in knowing how many frames and boxes to give the hive, when to add and remove them, and recognizing what is on the frames. We will be explaining this in the following chapters.

Right An illustration showing the different components of the hive, and how it works.

Below Movable frame hives allow you to remove the frames of honey from the hive without damaging the colony.

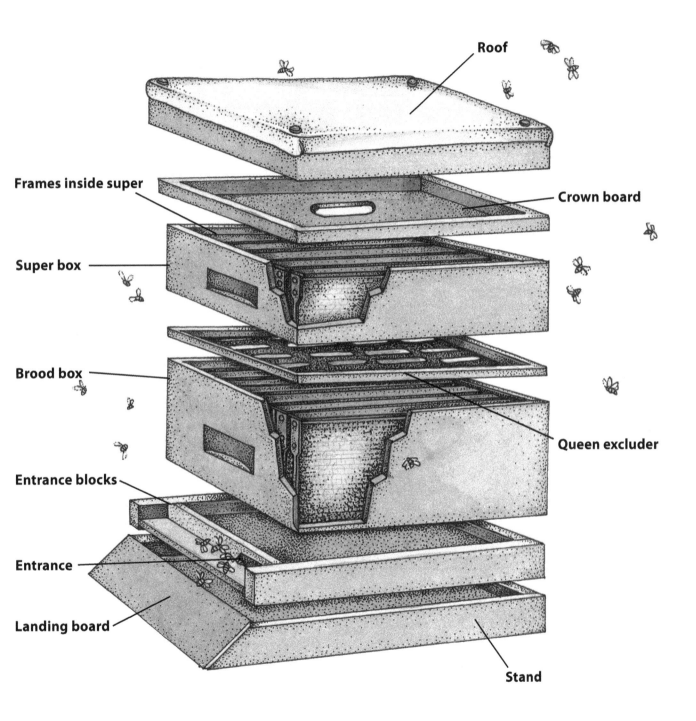

Roof

Frames inside super

Crown board

Super box

Brood box

Queen excluder

Entrance blocks

Entrance

Landing board

Stand

Different types of hives

The double-walled William Broughton Carr (WBC) hive

Ask most people to draw a picture of a beehive and they will draw what we call a WBC hive (pictured). William Broughton Carr was a keen amateur apiarist at the turn of the 20th century and designed a hive to suit the climate of the British Isles. With its gabled roof and sloping sides, it conjures up images of apple blossom, buttercups, gently flowing rivers and willow trees swaying in the breeze. A professional beekeeper, or even a keen amateur, will tell you that the WBC hive is expensive and unwieldy. However, it is the most attractive hive, so you will have to weigh aesthetic considerations against practicalities. Does adding such an attractive feature to your garden outweigh the disadvantage of any extra work involved?

Some of that extra work comes from the fact that the WBC is double walled. The outer parts (lifts), which we see as the sloping sides, act principally as insulation and as a casing for the inner parts. They have to be completely removed before you can get to the inside brood boxes and the supers. This does involve a bit more time and effort. On the other hand, since the brood and supers of a WBC are protected by the lifts, they are built with thinner wood than other, exposed hives. This reduces the weight of the inner boxes, which can be seen as an advantage.

The floor which both structures sit on acts as the stand and entrance with an attached landing board. The floor can also accommodate a Varroa screen for use in an integrated pest-management scheme (see 'Prevention and cure').

The bees' entrance to the WBC hive has a porch which keeps off the rain and accommodates the entrance blocks. You can easily slide these blocks open or closed to change the size of the entrance – you may want to make it smaller in the winter to stop mice getting in, or larger when the weather is hot to help ventilation.

The attractive gabled roof completes the hive. It contains ventilation holes and, since it overhangs the structure, it gives the hive that extra bit of protection against the elements by keeping rainwater off the lifts.

Above *The Commercial hive and, right, the Langstroth. Although they may look similar, they have their own bespoke parts that aren't interchangeable.*

Like other hives the WBC can be built from most wood materials, but cedarwood is recommended. Cedar's built-in oils and preservatives make it a great wood for the outdoors, as it can last decades even in extreme climates. It used to be the fashion to paint the outside of the hives white, but nowadays it is common to see the cedarwood left in its natural state. The bees don't mind either way.

Single-walled box hives

Single-walled box hives are by far the most common type used worldwide by professionals and hobbyists, and the components are easily bought from beekeeping suppliers. There are a number of designs and each region has its favourite. In some parts of the world there will be little choice on offer; in others you will find that there is a selection of four or five. The important thing to remember is that the component parts of different hives are unique to that design, so if you are buying extra frames or boxes, make sure that they are compatible with what you have already.

Box hives aren't as attractive as the WBC – in fact they look like a stack of boxes – but they are easier to use and the frames have more surface area, which accommodates more bees. For this reason, new beekeepers may choose to forgo the aesthetics and start with a single-walled box hive.

In practical terms, the style of hive you start with is likely to be the style you continue to use since you can't interchange the parts of different hives.

Box hives all work on the same principle – stackable floorless boxes which hold the frames – and look very similar, but they vary slightly in size. The single-walled version is made with thicker wood parts than the WBC: since the box is open to the elements, the wood needs to be thick enough to offer some degree of insulation from both the cold and the heat. The roofs and floors are separate items and the hives require a stand to keep them off the ground.

There are at least six different commonly used styles of box hives. This relatively high number comes down to eager enthusiasts with the time and money to devise new designs, and commercial chaps intent on making the most efficient honey-producing hives. Those who were successful in marketing their modifications gained a degree of following, which has led to some national preferences. The Langstroth is the most commonly used hive in the USA, the National in the UK and the Dadant in France. That is not to say that the Commercial, WBC and Smith are not used, but they are less popular.

The Dadant and Commercial are the largest and heaviest of these hives and, as the name of the latter suggests, are usually left to the professional apiarist. Our advice to the hobbyist is to use the lighter, more manageable hives.

Remember the 'bee space' (see 'How a hive works')? Well, some hives are 'top' bee space and some are 'bottom' bee space. The National, for example, is a bottom bee space while the Langstroth is a top. This means the space for the bees to crawl through to get around the hive is either below the frames or above the frames. This makes no practical difference to the workings of the hive, but it does highlight the fact that the various hive types are not interchangeable.

Above *The UK's most popular hive – the National.*

Other hives

Top bar hives (above) are used in many parts of Africa, where they can be built using cheap local materials. They have a movable frame but rather than using a constructed frame with foundation, the bees are encouraged to build their own honeycomb on bars, which are removed completely when the time comes to harvest the wax and honey.

A variation of the top bar hive is the long deep hive. Using the same frames as the single-walled hives, this design has been developed with the beekeeper – rather than the bees – in mind. Instead of growing vertically, the hive expands horizontally. It can be mounted higher off the ground so the beekeeper doesn't need to bend over so much and is designed to minimize the lifting of heavy supers.

tip
Nowadays it is common to see hive cedarwood left in its natural state. The bees don't mind what colour it is!

Hive components

The stand
A good sturdy stand keeps the hive off the ground. This stops damp and unwanted creatures creeping in and prevents the entrance being blocked by grass. But its main purpose is to raise the hive to a manageable working level. An old, small, low wooden table is a good start, as long as it can take the weight of the hive with a brood box and maybe four supers full of honey. About 60–90cm (2–3ft) off the ground should be enough; any more than that and by the time you add the fourth super the hive may be taller than you are.

The floor
This gives protection to the underside of the hive and also accommodates the entrance block, the movable piece of wood used to change the size of the entrance.

The Varroa screen
A piece of mesh that sits between the bottom of the brood box and the hive floor, the Varroa screen's purpose is to let the Varroa mite – a bee parasite – fall through the screen to the hive floor and eventually die. We talk about this in more detail 'Prevention and cure'.

The frames
These are the inserts that sit in the boxes of the hive and contain the wax foundation that encourages the bees to make the comb. Frames are either deep or shallow, for use in the brood and super boxes respectively. Those in the super hold the honeycomb and are the

ones taken out of the hive at harvest time. There are a variety of frame types that allow experienced beekeepers finer control over honey production. For the first year self-spacing frames are probably the best bet.

The foundation

Foundation is a pre-made sheet of wax moulded on both sides with impressions of hexagons. This gives the bees a start for their wax cells. They 'draw out' the foundation on both sides of the sheet by adding wax to the impressions, making the walls of the cells.

Foundation can be wired or unwired. The wired version is stronger and can be reused after the honey has been extracted. Unwired foundation is good for comb honey. A sheet of deep foundation for a National hive has 7,500 cells.

The brood box

The brood or deep box (the terms are interchangeable) is where the queen and the majority of the colony live. It is a busy place with the queen laying her eggs, larvae demanding food, newborns finding their feet, house bees cleaning the used cells, pollen and honey stored for food, and it all has to be kept at 35°C (95°F) by the bees vibrating their wings to generate the heat. It's dark, hot and busy.

Depending on the type of hive, the brood box holds between 10 and 14 frames and is around 23cm (9in) deep.

The queen excluder

This is a wire-mesh screen that allows worker bees to pass between the brood box and the super box above, but prevents the larger queen from moving between the two and laying her eggs in the super. The beekeeper wants the super solely for honey storage.

The supers

These are the boxes that hold super (shallow) frames with foundation. Each one holds between 10 and 14 frames and is around 15cm (6in) deep. This is where the bees store the honey that you will eventually harvest. In the course of a summer, you will probably need to add a number of supers to the hive to provide adequate storage space.

The crown board

A cover to fit over the top of the brood and super boxes, the crown board's primary purpose is to keep the heat in, so that the bees can maintain the temperature needed for the growing brood. It is usually made of wood, but there are glass and Perspex varieties which allow you to observe a little of what is happening in the hive without disturbing the micro-climate. In winter, glass or Perspex crown boards should be replaced with wooden ones which hold in the heat better. The cover board has a hole in its centre so that you can feed the bees without removing it.

The roof

The roof protects the hive from the elements and has ventilation holes that allow it to breathe. It is usually covered with some thin sheet metal for longevity. A WBC roof is slanted; single-walled box hive models all have a standard flat roof, but you can buy slanted ones to make them look more attractive.

Top *The Varroa screen is a piece of mesh that sits between the bottom of the brood box and the hive floor. It enables the varroa mite fall to the hive floor and die.*

Middle *Frames contain the wax foundation which the bees draw out to make comb. Between 10–14 frames sit in a brood box or super.*

Bottom *The crown board sits on top of the brood and supers to keep the heat in.*

How to make a frame with foundation

The principle for putting together a frame is the same for the brood and the super frames, since the only difference is their depth. The frame consists of a top bar, two side bars, two bottom bars and the nails to hold them all together.

The top bar has a holding bar that needs to be removed before frame assembly. It will be attached later to hold in the foundation. You can use the hive tool to remove this holding bar, but be careful – the sharp edge can easily cut a finger.

Once you have removed the holding bar, attach the two side bars to the top bar (this is a tight fit) with the grooves facing inwards. Place a bottom bar into the side bars (on the

same side of the frame as the thicker side of the top bar). Square the frame up and nail these pieces together.

The frame is now ready for foundation. If you are using wired foundation, bend the three top loops to 90 degrees and the two bottom loops back on to themselves so that they won't protrude from the bottom of the finished frame.

Starting with the edge which has the three loops, insert the foundation into the grooves of the side bars, making sure it stays flat, left.

Once the three hoops are touching the top bar, replace the wedge bar to hold the foundation and hammer three nails diagonally through the holding bar and the wire loops (left).

For unwired foundation just insert the foundation into the frames and hold it in place with the wedge bar and three nails.

The bottom of the foundation should sit on the bottom bar but not protrude out of it.

Fit the second bottom bar into the side bars and nail from the bottom. Ensure that the frames are straight.

tip
The principle for putting together a frame is the same for the brood and the super frames.

Right and Above Putting together a frame is a simple task, and is the same principle for the brood and super frames – the only difference being their depth.

What you will need

You will enjoy beekeeping when you feel comfortable and safe around your bees. For this, protective clothing is a must, along with certain pieces of equipment including a smoker and a hive tool. There is a wide range of protective clothing available, from all-in-one suits to smocks and veils. The suits are more expensive but do offer better protection. As well as traditional white cotton, they now come in lighter materials and other colours. Some manufacturers even offer a made-to-measure service. You will also need to think about honey-harvesting equipment. Manual honey extractors are adequate for the hobbyist with a few hives. An electric extractor is a pricier option.

The kit

Protective clothing

Looking after your bees will be stress-free and relaxing when you feel safe from the possibility of being stung. There is a variety of protective clothing available from beekeeping suppliers but whatever you choose to wear the idea is to present yourself to the bees in the least threatening way and to keep your skin covered.

The best way to achieve this is to kit yourself out with a full body suit. These all-in-one suits may make you look like a spaceman, but they do provide total protection, leaving you confident to get on with the work in hand. The basic type comes in white and is made of cotton, zipped at the front and with a veiled hood attached.

Alternatively, you can buy a suit made primarily from lightweight olive-coloured polyester. Its veiled hood can be unzipped at the front so that when you lift it over your head it lies comfortably at the back. (With some other suits you have to take the hood off completely, or else it dangles around your chest, getting in the way.) This design is not the cheapest on the market, but is by far the most comfortable, the least conspicuous and, importantly, the coolest. It can get hot out there inspecting your hive in the middle of a summer's day. Suits come in different sizes for adults and children, and some manufacturers offer a tailoring service at no extra cost.

The price for all-in-one suits varies between manufacturers, so do shop around. They are certainly worth the investment. A cheaper option is to buy a combined hat and veil. These incorporate a hat, netting and neck and shoulder coverings. The material can be tucked into a shirt or jumper, giving a great degree of protection without inhibiting your vision. In addition you can buy jackets or smocks that can be tucked into trousers, which are also sold separately.

The reason bee wear is light-coloured and smooth in texture is that bees don't like bears – big, dark, furry animals who steal their honey. So it is unwise to wear dark, woolly jumpers around bees, as this awakens their primeval instincts. Bees also like to climb upwards into dark spaces, so leaving your ankles and cuffs open and loose is inviting one to come and explore. It could then find itself trapped, stressed, under pressure and near to your flesh. This is likely to leave you with a barbed sting pumping a dose of histamine into your body, which is best avoided. Don't forget to cover your ankles with socks; stings on the ankle are memorable!

Another part of the body that needs protection is the hands. The best bee gloves are made from a good-quality goatskin. Although they offer complete protection, with long wrist coverings and thin leather, they are hard to clean and you do lose finesse. There are many beekeeping situations in which you will need to handle things gently or there are fiddly manoeuvres to accomplish. Common household rubber washing-up gloves protect the hands well while allowing you to maintain a delicate touch.

Some very experienced beekeepers are happy using their bare hands. If the bees are calm and your movements are gentle, you may get away without being stung.

Right Bee suits come in lightweight olive-coloured polyester, as well as the more traditional, heavier, cotton.

tip

A full body suit may make you look like a spaceman, but it does provide total protection.

The smoker

The image of a traditional beekeeper is of a bearded man wearing a veil and holding a strange contraption called a smoker. While the beard may be in retreat as more women and younger people take up the hobby, the Victorian-looking smoker is still an important piece of equipment. To understand why you use a smoker, you need to get inside the brains of your bees.

Bees can protect themselves against living creatures by using their sting and venom, but since a common natural habitat of a honey bee is a hollowed out tree there is another threat that presents a great danger and against which it can't use its sting: fire.

Bees are alerted to fire by the smoke that precedes the flames. Their instinct is to vacate the hive, get to a safe place and start a new home. But before they leave they spend as much time as they can collecting their worldly possessions, which in their case is the precious honey that they have been so busy storing.

HOW TO USE A SMOKER

The smoker is designed to produce cold smoke. It is not meant to erupt and spew out flames or heat. The bees just need to get a hint of smoke to believe there is danger about. The fuel that is placed inside the smoker is meant to smoulder rather than burn. This is why its canister has only a small hole at the bottom, which restricts the amount of air getting into it. The bellows are there to boost the airflow, which will increase the heat and set light to a bit more of the combustible material inside, allowing it to smoulder more and producing the cool smoke you desire.

So what type of material can you use that will produce lots of smoke, without too much heat, and will continue to smoulder and not go out just at the moment you need to calm the bees down?

Hay, grass cuttings and wood shavings all do the job. Cardboard works very well when rolled up into a cylinder with the veins running vertically, creating an airflow. Combining cardboard with hessian sacking – the type that is used for sand bags and vegetables – makes an excellent fuel that can be made up in advance. Hessian is not as common as it used to be before the advent of synthetic materials, but it is still available from beekeeping suppliers. Cut both hessian and cardboard to the height of the smoker and roll them up together, a layer of hessian wrapped in a layer of cardboard. Light the material, push it into the smoker, pump the bellows a couple of times and you should have a working smoker that will last 30 minutes or so.

As with any material you propose to use in a smoker, ensure that it has not been treated with fire-retardant chemicals, as is sometimes the case with cardboard packaging. The bees will not take kindly to the noxious smoke produced by this type of fuel.

A gentle puff into the entrance five minutes before you open the hive should give the bees enough time to get them busy sucking honey. They will generally move away from smoke, so the smoker can also be used to herd them off a frame. Too much smoke, however, can make them angry and likely to sting.

Using your smoker effectively is an important aspect of the art of beekeeping. It can be a bit tricky at first, so it is worth spending time practising until you get it right.

Clip Queen Catcher

Press-in Cage

Bee brush

Cold Uncapping Knife

Plastic Porter Bee Escape

Smoker Cartridge

Hive Tool and
Frame Lifter

Above *Some of the pieces of essential equipment that help beekeepers with their craft.*

Diving into the comb, they quickly suck up enough honey to enable them to survive for three days without collecting more while they are setting up their new home. But just as we get soporific when we have gorged on a big meal, so bees with their full bellies become less aggressive.

This is why the smoker is used. A little blast from its snout will give the bees enough of a fright to make them busy themselves while you get on with your work in the hive. By the time they realize that it was a false alarm and get back to their routine, with a bit of luck you will have finished disturbing them.

Hive tool

Another essential piece of beekeeping equipment is the hive tool. This is basically a shaped piece of metal sharpened flat at one end and hooked at the other. The sharper end is used as a scraper and the blunt, curved end as a lever.

The main reason you need this tool is that bees stick most parts of the hive together with a substance called propolis (see 'Bee species') and it gets everywhere. The frames you put into the hive are stuck together with small amounts of propolis, the super is stuck to the box below it with propolis, the crown board is stuck to the supers – and so on and so on. The hive tool is invaluable for gently prising the parts away from each other. Before long you

will find that whenever you are near a hive you'll feel bare without the hive tool firmly in your grasp. It also comes in handy when making frames, since it works perfectly for separating the holding bar from the top frames (see 'How to make a frame with foundation').

Skeps

Bees have a tremendous sense of location and if two hives are close together they know which entrance is theirs. On occasion it is necessary to move the hive away from its usual location temporarily, but the bees will still fly back to where the entrance was. To keep those bees safe and happy until you return their home, you need a skep. This gives year you may need to feed them. In the late autumn after you have harvested much of their honey, feed them a strong solution of sugar and water to help see them through the cold winter months. In the spring, feed them again but use a weaker solution. This lets the bees know that the nectar is back in flow and encourages the queen to increase her laying.

How to feed bees

As with most aspects of beekeeping, there is more than one way to do it. The easiest and most common type of feeder is a rapid feeder, which has the same outside dimensions as the brood box and sits on top of it. It has an opening where the bees can come up into it from below, and a well that contains 4.5 litres (1 gallon) of liquid. Many bees can get to the sugary solution, which may be gone in a matter of days. It is clean and easy to use, as the liquid can be added when the feeder is in place on the brood box.

The other type is a contact feeder – basically a plastic bucket with a tight lid and a mesh-covered hole. When the bucket is filled and turned upside down, a vacuum is formed so that the liquid stays put, but the bees can get to it through the mesh. Contact feeders should be placed over the holes on the crown board and the roof supported on an empty box. The most common versions hold either 1.1, 2.2 or 4.5 litres (2, 4 or 8 pints) of liquid.

Honey harvest

We will go into the details of extracting honey later (see 'How to harvest honey'), but the equipment you will need includes an uncapping knife or fork, bowls, jars, sieves, an extractor – either manual or electric – and buckets for containing the honey. The process can get costly if you buy electric extracting devices. You may be able to borrow an

Above Skeps were once used as full-time hives. Their main purpose today is to act as a temporary home in case the hive needs moving.

Right Harvesting honey requires its own equipment, including an extractor.

them a dark place to shelter while you get on with what you are doing. A traditional skep is a deep bowl-shaped container made from woven straw. You can use a cardboard box – it doesn't look nearly so nice but it serves the same purpose. Once the hive is returned to its place the bees in the skep can be shaken into the hive.

Feeders

Bees don't rely on humans for their daily food, but two or three times during the course of a

extractor from your local beekeepers' association, but at busy times of the year there is likely to be a waiting list. You also need to think about where to do your honey extraction. It needs to be bee-tight (that is, with no opening that bees can get through), since they will smell the honey from a long way away and tell their friends about it. Before long you'll have bees buzzing all around you. Your kitchen will be fine, but the extracting equipment will take up most of the space, so it is unlikely you'll be able to make dinner at the same time. Honey harvesting is a sticky business, so cover any surfaces you are using with newspaper or an old tablecloth.

Where and when to get your bees

Once you have set up and located the hive, bought your suit and equipment, and have a basic understanding of how bees live and work, you will need to find a supplier of good-quality bees. Many suppliers and commercial farmers advertise in beekeeping magazines, but where possible go for word-of- mouth recommendations. Alternatively, your beekeepers' association may be able to provide you with a swarm. In either case, early summer is the best time for new bees. You will need to know in advance how many bees you are getting and what type they are.

Getting your bees

Knowing a beekeeper

Depending on the demand for bees in your area you may be offered a swarm of bees from a local beekeeper for no cost. But in some areas the demand for bees is high and you may well need to spend money on obtaining your colony. The choice of where to get your bees will usually come down to a friendly beekeeper, or finding a club or association or a commercial beekeeper.

Knowing a beekeeper who has a spare colony is often the best way for a beginner to get started. The beekeeper will generally help you and make sure you are on the right track.

Beekeepers' associations

Beekeeping associations and clubs are a good source of bees and they often offer theory courses during the winter and mentoring programmes in early spring and summer which will give you time to handle live bees before you start with a colony of your own. The 'swarm coordinator' gets the call to pick up swarms from the local neighbourhood. They pass the swarms to eager members waiting for their bees, from beginner beekeepers getting started to more experienced beekeepers expanding their apiary.

The disadvantage of a swarm is that you may be waiting a long time for one. You may be keen to get going with your own bees in the spring but natural swarms can happen at any time so it is hard to plan for them. If the weather is wet and cold, the bees will be late swarming and it may be well into the season before you can set up your own colony. Even if the weather is good and there is high demand for swarms at your local association, your name may be low down on the swarm list and others may get the only available swarms that season.

Commercial farmers

A more reliable source of bees is a commercial beekeeper. They may be honey producers, or migratory pollinators moving their beehives around to pollinate crops that are dependent on bee pollination for larger yields, or they could be raising and selling bee colonies. Most of them will be able to sell you a nucleus. Your local beekeeping association may know of local commercial operators, and beekeeping retailers will certainly be able to point you towards a source of live bees, even if they don't sell them themselves. The journals and magazines dedicated to the industry carry adverts for bee suppliers.

Wherever you get your bees, the main thing to know is that there is a small window of opportunity to acquire bees for your new hive and most of the changing of hands is done in the spring and early summer. This is due to the life cycle of the bees. It is in the spring that they begin to increase the size of the colony so expect to get your bees at that time of year. It is best to get in touch with a supplier sometime in the winter and place an order. Supplies are limited since demand for bees is on the increase.

Left *Whether you buy a nucleus of bees or collect a swarm, they will come in what's called a nuc box – a travel hive that is used for transporting a small colony.*

WORLDWIDE BEE ORGANISATIONS

THE AMERICAN BEEKEEPING FEDERATION acts on behalf of the US beekeeping industry on issues affecting the interests and the economic viability of the various sectors of the industry. It keeps tabs on what's happening in Washington that might be of interest to beekeepers and helps them to promote honey with promotional material and funds honey purity research. It operates as a resource for and about the US beekeeping industry, hosts conventions and shows, and provides members with up-to-date information via a bi-monthly newsletter. Importantly, it provides liability insurance for local and state-wide beekeepers associations. **www.abfnet.org**

THE BRITISH BEEKEEPERS ASSOCIATION (BBKA) exists to promote beekeeping as a hobby and a profession. One of its aims is to encourage new people into this fascinating craft by providing contact with experienced beekeepers who can give a helping hand, running introductory courses, and putting them in touch with a local association. British beekeepers usually join a local association. Part of their membership fee includes membership of the BBKA which entitles them to a bi-monthly newsletter, public and products liability insurance as well as insurance for brood diseases. The BBKA also conducts examinations in beekeeping. **www. bbka.org.uk**

THE INTERNATIONAL BEE RESEARCH ASSOCIATION (IBRA) is a Welsh-based charity, founded by the late Dr Eva Crane, which works to increase awareness of the vital role of bees in the environment and to promote beekeeping as a sustainable economic activity in developing countries. It publishes the highly respected quarterly *Journal of Apicultural Research* for scientists and beekeepers worldwide. **www.ibra. org.uk**

NATIONAL BEE UNIT
The UK government-funded National Bee Unit delivers the bee health programmes on behalf of the Department for Environment Food and Rural Affairs (Defra) in England and the Welsh Assembly Government (WAG) in Wales. Its function is to protect the honey bee, an essential pollinator of crops and wild plants, from threats such as serious disease and environmental damage. It does this by using apiary inspections, diagnosis, research and development, and training provided by a team of bee scientists and experienced practical beekeepers. It provides specialist advice and technical consultancy services on bee health to UK government departments, beekeepers and industry. **www.nationalbeeunit.co.uk**

THE FEDERATION OF IRISH BEEKEEPERS' ASSOCIATIONS promotes beekeeping in Ireland and is a federation of all the local associations across the country. It provides information for novice and experienced beekeepers on bee health and management and runs summer schools, conferences and honey shows. **www.irishbeekeeping.ie**

AUSTRALIAN HONEY BEE INDUSTRY COUNCIL is an industry body whose mission is to ensure the long term economic viability, security and prosperity of the Australian honey bee industry by lobbying government, private business and the public in the interests of the industry and conducting educational, promotional and public relations campaigns. **www.honeybee.org.au**

NATIONAL BEEKEEPERS' ASSOCIATION OF NEW ZEALAND provides information and advice for hobbyist and professional beekeepers on the north and south islands, and puts them in touch with their local association. It publishes the New Zealand beekeepers' magazine and educational material, and organises events throughout the year. **www. nba.org.nz**

CANADIAN ASSOCIATION OF PROFESSIONAL APICULTURISTS is a bi-lingual association that is open only to people who are employed in apiculture, either as academics or through government organisations as researchers, regulators, or bee inspectors. It disseminates information about beekeeping, co-administers the Canadian Bee Research Fund and develops research priorities.
www.capabees.ca

ISTITUTO NAZIONALE DI APICOLTURA was founded in 1931 at the Department of Zooculture, Faculty of Agriculture of the University of Bologna and funded by the Ministry of Agriculture and Forests. The institute has become the main point of reference in Italy for problems and issues related to beekeeping. The institute's labs were the first in Italy to specialise in the instrumental insemination of queen bees and to apply this technique in programmes for the genetic selection of honey bees. It has established partnerships with beekeepers' associations in the management of disease prevention plans and provision of diagnostic services.
www.inapicoltura.org

THE AFRICAN BEEKEEPING RESOURCE CENTRE aims to assist the development of beekeeping in Africa as a means for people to generate sustainable livelihoods and beat poverty by sharing knowledge and ideas, and creating linkages and friendships. It believes that knowledge-sharing is a two-way process and that beekeepers in other parts of the world can learn from the wealth of traditional African beekeeping knowledge. African bees can be aggressive and require appropriate handling and management. Africanized bees have spread from Brazil where 26 colonies were imported in 1956 and are now expanding in the southern United States of America.
www.apiconsult.com

PRACTICAL ACTION SOUTHERN AFRICA is a South African-based voluntary organisation of both hobbyist and professional beekeepers. Hobbyist beekeepers may have one or two hives for pure pleasure and recreational interest or up to 50 hives for a more lucrative pastime. Professional members may have many hundreds of hives for full-time honey production and pollination services. It promotes and advances the science of apiculture, maintains competency and professional conduct among its members, assists in the establishment and maintenance of educational and training facilities, promotes the interests of the beekeeping fraternity, and publishes a quarterly newsletter. **www.practicalaction.org**

UNION NATIONALE DE L'APICULTURE FRANCAISE was created more than 60 years ago by a handful of beekeepers, most of them professionals, to defend their interests against the government. It now represents 100 regional groups and about 22,000 beekeepers. It has fought against unfair competition, is at the forefront of banning pesticides and has helped to develop cures for bee diseases. It publishes a monthly magazine for commercial and hobby bee keepers. In 2005 the organisation launched a project which aims to keep the public informed of the damage of pesticides to the environment. **www.unaf-apiculture.info**

What you get

There are three ways of purchasing a colony:

- as a nucleus of bees, which is 10,000 workers, a laying queen, brood in all stages of development, and stores of pollen and honey on five frames which comes in a nuc box
- as a package of bees with 10,000 workers and a laying queen but without frames of brood, pollen or honey
- as a full colony of 50,000 bees supplied in a working hive

A free swarm of bees is similar to a package of bees as it doesn't come with any brood, stores or frames. It may be a bit bigger in size, containing up to 20,000 worker bees.

Nucleus

A nucleus of bees is a developing colony of around 10,000 working bees, one queen and some drones all living on the wax honeycomb that they have built and stuffed with stores of honey and pollen to feed the brood. The nucleus should be five frames with at least three frames of brood in all stages so there will be eggs, larvae and pupae, one queen, who is the mother of all the bees within the nuc, and two frames of stores of pollen and nectar. There should be bees on all of the frames so that the nucleus looks full and packed with worker bees.

Beware that some suppliers have been selling their customers short with nucs that have too few bees and brood. Try to get a recommendation for a commercial farmer from a trusted source. Before placing your order there are a few questions you should always ask regarding the nucleus:

1. Do at least three frames have brood in all stages of development?
2. Is the queen the mother of all the bees in the nuc? (Sometimes the supplier will introduce a queen who isn't the mother. This isn't necessarily a problem as the worker bees should accept a new queen after a few days, but mum is best.)
3. Has the colony been treated for the Varroa mite and is it free from disease?
4. Are the size of the frames going to match the frames in your hive? (You will also need to stipulate the type of hive that you are using since the frames are different sizes for different hive types, but a good commercial supplier will be able to sell you a nucleus that fits your hive.)
5. Does the seller offer a guarantee to replace the colony or queen if things go wrong, or if disease is found? You can also ask for the queen to be marked with the appropriate colour for her age. Most bee farmers will do that for you.

It will be difficult for a farmer to give you an exact date when the colony will be ready since this depends on the weather and how well the colony has lived over the winter, so you will need a degree of flexibility when it comes to the date to pick up your bees.

If you take delivery of your colony in the spring or early summer you will need to transfer them into the new hive as soon as possible as the colony will want to expand and the queen will need the extra room to lay her eggs.

More than likely you will drive to pick up your bees but in some cases you can get the bees delivered – check with your supplier. It can get hot in the back of a car for the bees in the nuc box, even if the box has a mesh ventilated roof for travelling purposes, so it is advisable to collect your bees in the early evening when it is cooler.

The advantage of buying a nucleus is that you are buying a small developing colony that is ready to expand. The colony's cycle is not interrupted and you can see that the queen is laying eggs and the brood is developing.

Package bees

Package bees are cheaper than buying a nucleus because you are only buying the bees and a queen without the brood or the frames. A package is often sold by weight with a 3lb package giving you approximately 10,000 bees.

You get a queen and 10,000 workers housed in a screened box with a tin of feed. The queen and the bees may be from different colonies so they need to get used to each other's smell. To achieve this, the queen is housed in a separate screened cage keeping her safe. This allows her smell to permeate through to the workers and she can still be fed through the screen. It takes a few days for the bees to accept her as if she was their own mother but once they have, they will all live happily together.

Package bees can be delivered by courier. Try and arrange the delivery to be made at a convenient time so the bees aren't in transit over a weekend. Picking the bees up locally is less stressful for them.

The disadvantage of package bees is that there is no developing brood. This means that it will be at least three weeks before the colony gets new young bees and a further three weeks until they can forage. Remember that the foragers only live three weeks so your small colony of package bees will reduce in population before they are replaced by the new young. You will need to feed these bees a sugar syrup solution (see 'Transferring your bees') so they have a plentiful source of available energy to make wax and feed the developing young.

The advantage of buying a package of bees is that the bees may have less Varroa because the mites breed in the brood nest. You are also reducing the risk of brood diseases.

Full colony

A full colony of bees is a working hive ready just to pop into position in your garden. This fast-track way of getting into beekeeping does have its advantages but there are some draw-backs too. You are buying second-hand equipment so it would be wise to get the hive checked by an experienced beekeeper to make sure it is free of disease and long-lasting. Moreover, taking charge of a fully working hive requires the sort of knowledge and management that is normally gained after the first year of beekeeping. For a novice, it is diving in

at the deep end. It could be pretty daunting taking a delivery of 50,000 bees, which is what a fully functioning hive holds at the height of the summer. Gaining confidence around just 10,000 bees takes time. It is also easier to see what is going on in the hive when you first inspect it if there are fewer bees. For both these reasons it is recommended that novice beekeepers start with a nucleus or a package/swarm and build up to a full-size colony.

tip

Being able to spot the queen in a hive is vital, because it is important to know that she has not left the colony.

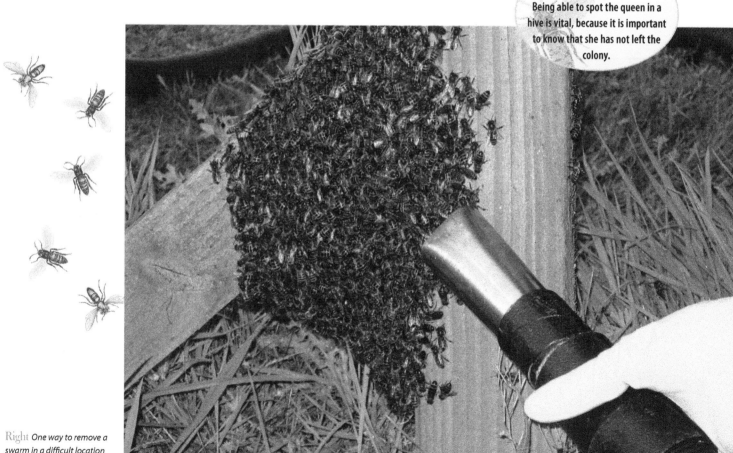

Right *One way to remove a swarm in a difficult location is to use a specially-adapted vacuum cleaner to literally suck them up.*

Swarm

A swarm of bees is similar to a package of bees in that you only get the adult bees and not the brood. You will probably get the swarm at very short notice as it will probably have just been collected by an experienced beekeeper. Be ready to drop everything to collect it, otherwise you could find it going to the next person who had put their name down on your association's swarm list.

If you are lucky and the swarm is a 'prime' swarm from a strong colony you will be getting an experienced mated queen who is the mother of the 20,000 or so bees. The bees' crop will be full of honey and they will be ready to build the new wax comb as soon as they are housed in their new home – your hive. The queen will start to lay in the new comb as soon as it is built and the colony will begin in earnest. If, however, you are unlucky and get a 'cast' swarm you will be getting an unmated queen and maybe only a few thousand worker bees. In this case you will need to hope your queen is mated fast and can begin to lay eggs in the new home. Ask the swarm collector if they know if it is a prime or a cast swarm. Either way you will soon have a new colony in your new hive. You may need to give them some sugar syrup to encourage them to build the wax.

Bee breeds

Whichever method you use to get your colony, you hope your bees will be gentle and well-behaved. Beekeepers generally want a docile bee that doesn't want to swarm too often and stores lots of honey.

Honey bees, managed by us, are all the same species; the western honey bee (*Apis millifera*) which originated in Africa and is indigenous to Europe. Within this species,

MARKING THE QUEEN

If you are buying a nucleus or package, it is worth paying a bit extra for the queen to be marked. Being able to spot the queen within the mass of bees in a hive is very useful to the beekeeper because it is important to know she has not left the colony, either through swarming or dying, and has not been damaged. Since you don't want to harm her accidentally when you are inspecting the colony, if you can find her, you can isolate the frame she is on while you safely examine the rest of the bees.

If your queen has not already been marked you can do it yourself with paint purchased from your beekeeping supplier. There are a number of different devices available to help with this, though experienced beekeepers use their fingers to carefully hold the queen while applying the paint. The principal is the same either way: you need to trap the queen and hold her still while you dab a drop of the paint on her thorax. Once the paint has dried release her to return to her colony.

The colours used for marking are internationally recognised and follow a pattern; blue, white, yellow, red and green for the years ending in 0 or 5, 1 or 6, 2 or 7, 3 or 8, and 4 or 9 respectively. Since the queen doesn't normally live longer than five years, this method allows you to tell her age at a glance.

Marking the queen may sound straight forward, but finding her in order to be able to apply the mark is easier said than done as she is usually hidden by hundreds of worker bees and shuns the light. This is why having a queen who is already marked is a good idea.

Above *A marked queen. If you are going to mark the queen yourself you need to isolate her with a queen press-in cage and then put a mark on her.*

however, there are many different races with different characteristics according to where they have evolved in the world and adapted to their surroundings. For example, the northern European honey bee (*apis mellifera mellifera*), often called the 'black bee', is more adapted to colder northern regions. It has dark, long hairs to protect it from the cold, is slow to develop brood in the spring when the weather can still be chilly, and it is also known for being more defensive than some other races, but it ekes out its honey stores during the cold winter months so is less likely to starve than other races which are less frugal eaters.

In contrast, the golden-coloured Italian honey bee (*Apis mellifera ligustica*) is quicker to expand in the spring, makes more honey and is a more gentle bee. It can, however, eat its winter stores too quickly and may be less resistant to pests, parasites and intemperate weather in colder climates.

The Italian bee has become a very popular choice around the world because of its prolific honey-making and its more docile nature.

Left *Bees coming out of the hive. They are a local hybrid so slightly different colours – some more golden, some darker.*

Above *Nuc boxes being picked up and transported.*

If you are buying your bees from a supplier ask what type of bees they sell. They often have a choice. Other favourites are the darker Carniolan bee (*Apis mellifera carnica*) which originated in the Balkans and the Buckfast bee, a crossbreed bred almost a century ago by Brother Adam, a monk at Buckfast Abbey, Devon.

Bee experts and scientists now recommend that you should buy a local hybrid bee because it is that thought they will be best suited to the local weather conditions and the flora and so better able to survive. Because queens mate on the wing with drones from different colonies around the local area, more often than not the colony you will get will be a hybrid – an Italian queen, for example, carrying the genes of the local drones.

Whichever bee you end up with you will want it to be gentle. Beekeeping is so much more enjoyable when you are not in fear of being stung. Aggressive bees will be noisy, the guard bees will follow you up the garden and try and sting you through your suit. It is thought to be the pheromones and genes from the queen that influence the behaviour of the bees. If you are unlucky enough to get these types of bees you will want to replace your queen with one from a more gentle stock. An experienced beekeeper will be able to tell you the different techniques for changing your queen.

The character of the colony may change over time, especially when the old queen has swarmed and the new queen has mated with local drones. Keep a record of the behaviour of your colony and re-queen if necessary.

Health

Whatever race of bee you get, it is vitally important that you set out on your beekeeping journey with healthy bees. Remember that the health of the colony is largely determined by the state of the brood.

As a novice, recognising the signs of ill health within a colony is not easy – even experienced beekeepers have difficulty in spotting signs of disease – so the best advice is to obtain your new colony from a trusted source and to get the colony checked by a more experienced beekeeper than yourself.

Making your bees feel at home

The bees in a newly purchased nucleus are likely to be tired and hungry from their journey. A 'welcome meal' will give them the energy to set up their new home, while watching and listening to them coming and going can tell you a lot about their wellbeing. The foragers will seek out the best sources of nectar and pollen and if their back legs are covered in pollen you know that they are busy bringing in food for the hive. Soon the colony will outgrow its living space. Inspect it after nine days and every seven days thereafter, to judge when to add an upstairs extension. As long as your bees have enough room during the summer, you don't disturb them too often and the weather is warm, they should build a strong and healthy colony.

SUGAR-SYRUP SOLUTION

Sugar-syrup solution is similar to nectar and is a good supplement to the bees' natural diet when food is scarce or when you want to energize them. To make the solution, simply add white cane sugar to warm water and stir until you are sure it has all dissolved. It is important to use cane sugar, since beet sugar upsets the bees' digestion.

The strength of the solution varies according to the reason for the feed. Give a strong solution in the autumn (see 'Preparing for winter') and a weaker solution in the spring and when you are hiving a new nucleus. A strong solution is a 2:1 ratio of sugar to water. A weak solution is a 1:1 ratio and it doesn't matter whether you measure in volume or weight. About 2.25 litres (4 pints) makes a good welcome meal for 10,000 bees.

The warmer the water the more quickly the sugar will dissolve, but let it cool before giving it to the bees.

Transferring your bees

Having arrived home with your bees, how do you get them into your hive? If you have a nucleus in a five-framed nuc box, the first thing to do is to place the box on the site where the hive will be and leave it for 24 hours. If you have had an empty hive in position awaiting the bees' arrival, move it away and place the nuc on the floor of that hive. This gives the bees a chance to settle down and get their bearings. They may be a bit hot and bamboozled. Don't worry if they sound angry – they'll soon cool down in the garden.

Unblock the entrance of the nuc box, but leave something like a clump of grass or a bit of leafy branch partly blocking the entrance. This forces the bees to take a bit more time on leaving the nuc, so they can readjust to their new location. The last time they left the nuc they were in a different place, so they will need to reorientate themselves. You don't want them darting out of the hive and getting lost.

While they are getting their bearings you can make up tomorrow's welcome meal of a sugar-syrup solution (see 'Sugar-syrup solution'), and practise lighting the smoker.

The evening after the bees arrive you can begin the transfer to the hive. This requires moving the frames containing the bees from the nuc box into the middle of the brood box of your new hive. Light your smoker (see 'The kit'), put on your protective gear and walk over to the nuc. As it is evening, most of the flying bees will have returned and will be in the nuc, though there will be a few hanging around the entrance. This is normal, especially if it's hot. Direct a light puff of smoke at the entrance and gently move the nuc box away from the hive site. The smoke will have time to work on the bees while you put the hive into position. Remove the roof and have an empty brood box ready in position on the hive floor.

Bring the nuc box close to your hive and, using your hive tool, gently prise open the box's travelling roof. Then use the hive tool to lever an end frame out the box – it will be stuck together with propolis. When propolis is warm it gives way under pressure, so when moving and lifting frames start with a little pressure and increase it when you feel the frames starting to separate. Don't jerk the frames apart, as that will alarm the bees.

Lift the first frame, with all the bees, from the nuc box and place it in the centre of the brood box. Do the same for the other frames but make sure that once they have been moved they are touching their neighbouring frames in the same way and position them as they were in the nuc box. If you are lucky enough to spot the queen, be extra careful with the frame she is on, since you don't want her to be damaged or to fall off on to the ground.

Once all five frames have been transferred, lots of bees will be left at the bottom of the nuc box. Gently turn it upside down over the brood box and give it a firm jolt to get most of the remaining bees

tip

Don't worry if the bees sound angry at first. They'll soon cool down in the garden.

out. Then lean the nuc against the entrance of the hive and the stragglers will eventually make their way inside.

Now, gently close up the five frames you have just transferred to the centre of the brood box. Fill the spaces on either side with new frames and foundation. Depending on the type of hive you have chosen, the brood box will hold 10–12 frames in total. When you are happy that the frames are tightly together and straight, place the cover board on top.

You have just got your first colony of bees into their new home! Before you put the roof on you need to give them that welcoming sugar-syrup solution. Because you have only

10,000 bees, a contact feeder is best here (a rapid feeder comes into its own when you have the full complement of 50,000). Pour the liquid into the bucket, replace the lid securely, and turn the feeder upside down. A little liquid will come out at first, but it will stop as soon as the vacuum is created in the bucket. Perform this operation over a bucket or a bit of ground away from the hive, so that the spillage doesn't attract robbing bees and wasps.

Once you have turned the feeder over, place it over the hole in the crown board. Now that the feeder is in place the roof won't fit, so you'll need to place an empty super on the brood box and then put the roof on that.

Above *When you get yout nuc box home you will need to put it in the final location of your hive, so that the bees can get their bearings. The frames need to be carefully transferred from the nuc box into the new hive.*

Reduce the entrance of the hive to an opening of 2.5–5cm (1–2in) so that the small colony can defend it. Leave the nuc box and lid leaning against the entrance in case stragglers need more time to find their way into the hive.

Leave the bees to settle into their new home overnight as you celebrate becoming the proud custodian of a colony of bees.

What's in the hive

The first few weeks after your bees arrive are really exciting as you get to understand and identify what is happening in the hive. Books can try to prepare you for the experience, but the sheer pleasure will only become apparent when you see and smell the hive for yourself.

The day after you've transferred your colony into the hive, remove the nuc box which will now be empty of bees. Spend a few minutes watching the entrance to the hive and observe the comings and goings. Some bees will be wandering around the landing board, some will be flying away to forage, others will be returning with brightly coloured pollen on their legs, or with nectar in their stomachs. There will be a few bees flying around the entrance. These are foragers orientating themselves before flying away.

Inside the hive the bees will have cleaned up any of the mess created by the first night's move. Cells that opened and leaked honey will be repaired and the bees will have begun to look around their new home. The queen will be laying and the house bees will have started to draw out the cells of wax on the new sheets of foundation.

You need to wait until the hexagonal frame on each cell is about 1cm (½in) deep before adding a super. Bees naturally work vertically, so if you add a super too soon they will leave the frames at the sides of the brood box empty and move upstairs. The more cells available to the queen for laying her eggs in, the bigger the colony and the better your chance of getting a good crop of honey.

Watch and listen to your bees as much as you can. You can learn a lot about what is going on in the hive this way. Lots of pollen coming in means a healthy laying queen with plenty of brood to feed. Bees standing on the landing board with their wings beating are busy circulating the air to evaporate the nectar to make honey. There should be a quiet humming sound coming from the hive.

When you first have bees it is tempting to open up the hive and have a look at them every day. But do that and they will probably head off to find a more peaceful place to live. So curb your enthusiasm. Wait nine days (that's the time it takes for the colony to cap the first batch of eggs) before looking inside.

First inspection

When the time for your first inspection comes, have everything ready before you start. It is a good idea to have a dedicated bee box or basket with all your equipment: smoker, fuel and matches; hive tool; gloves and protective gear; notebook; and a white sheet to lay on the ground or cover the hive.

Pick a warm day and plan to look at the hive in the early afternoon when the foraging bees will be out. This means that there will be fewer bees in the hive, which will give you a clearer view of the frames. A warm day also helps the bees maintain the temperature of the hive. Remember that when you open the hive all the hot air is lost and replaced with cold air. They can handle this for a short period so act quickly but gently.

For this first inspection you are going to look through each frame and

ascertain what's on it. You are hoping to see eggs, larvae, sealed brood, pollen, honey and sealed honey. If the queen is marked you may catch sight of her, but don't worry about identifying her at this stage. As long as you see eggs, larvae and sealed brood you know she is there somewhere, busily laying. Look out also for the distinctive queen cells (acorn-sized cells which house a queen in the making) which show that the colony is thinking of swarming. We talk about these and their effects in 'Swarming and increasing an apiary'.

A mental run-through of what you are going to do, envisaging where you will stand, where you will place the items you are moving and what you are looking for, is a valuable aid in the preparations. For a beginner, it is also comforting to have someone else to bounce ideas off and to learn from. If one of you forgets something you have read in a book, chances are the other will remember. From a practical point of view, having two people opening up and going through a hive together can be easier than doing it alone. One can hold up the frames while the other says what they can see and if you want to keep a visual record of your inspection one of you can take photos. It's also great to be able to share your excitement with someone. If you are a lone, novice beekeeper you may want to consider asking a member of your association to take you through your first inspection.

Getting started

Light your smoker just before you put on your protective clothing – if you try to blow out a lighted match with your

veil on there is a good chance you'll burn the veil. Then put on your gear, pick up the hive tool and approach the hive from the back or side. If you stand at the front of the hive you will be blocking the bees' entrance and this will annoy them.

Take the smoker to the entrance and give it a gentle puff or two. Wait for five minutes to let the alarm call spread through the hive. While that is happening you can double-check that you have everything and that the space around the hive is clear.

Lift off the roof. It should come away easily, but if the bees have used some propolis to fill a gap you may find that it sticks for a moment. Remember to use only gentle pressure.

Have a look at the underside of the roof. There may be a few bees walking around the surface. Can you tell if they are workers? Is there a drone? If you spot the queen, place the corner of the roof next to the hole on the crown board and gently encourage her to go down into the dark hole. She will be very

tip
Light your smoker before you put your veil on – trying to blow out a match with it on could lead to damage.

happy to do that. However, she will almost certainly not be there because she will be busy laying on one of the frames. Place the roof upside down on the ground to the side of the hive.

Remove the empty super you used to house the feeder and place it to the side of the hive. Take the feeder off the crown board. Note whether the feeder is empty. If it isn't, you can give it back to the bees after the inspection. Look down through the feeder hole of the crown board and you should see

some bees. (You can give a small puff of smoke through the hole, but this isn't necessary if the bees are calm.) The crown board will certainly be stuck with propolis, so run the sharp end of the hive tool around the underside of the crown board to break the seal and prise off the board. Try to be gentle and quiet: the bees will hardly know you are there if you work smoothly.

There could be a lot of bees on the underside of the crown board. This is perfectly normal. You should clean any comb off before returning the crown board, but for the time being lean the board against the entrance. This will give the bees on it a chance to walk back into the hive.

Inspecting the frames

Now you are going to inspect the individual frames. Use the hive tool as a lever to break them out of position. Start from one end and, gently using the curved end of the hive tool, lift the frame out. You may find that it is a frame of foundation and nothing else. That's fine. Most of the action in the hive will probably be nearer the centre, since the bees keep the brood nest together in the shape of a football and they won't have had enough time to expand to the edges. Place the first frame inside the upturned lid on the ground. Now the second frame. Again you may need

> Most of the action in the hive will probably be nearer the centre, since the bees keep the brood nest together in the shape of a football.

the hive tool to prise it along. Slide the frame away from its neighbour before lifting it out to save crushing any bees. Is this frame any different? Does it look as if the foundation has some new wax on it? Maybe your bees have begun drawing out the wax, making new cells. Or they may not have got this far yet.

Place this frame into the box on the ground too. This will give you more room to slide out the other frames. Continue sliding each frame away from its neighbour before lifting it out of the hive. Keep the frames over the hive when you are inspecting them in order to catch any bees that may fall off. As you get closer to the centre of the brood and to the frames you transferred from the nuc box you will find more bees and more activity. You should notice some cells filled with liquid nectar and others with different-coloured pollen.

Some cells will have a white wax covering which is stored honey, others a darker coloured wax covering which is sealed brood. What you want to see are some cells with larvae at different stages. The bigger, pearly white, 'C'-shaped larvae are quite easy to spot; next to them you may notice smaller larvae and possibly, at the bottom of the cell, a tiny white egg. To get the best view, have the sun over your shoulder so that the sunlight falls into the cells.

If the frame you are holding has eggs on it, there is a fair chance that it will also have the queen. She tends to move around the frames quite quickly. You will recognize her as soon as

Above and Right
On your first inspection of a frame in the centre of the hive, you will want to see brown capped cells (sealed brood), pearly-white larvae, pollen and white capped honey cells.

you see her from her long body and legs. Once seen never forgotten. If she is on that frame, take extra care not to damage her as you replace it. You may want to place the frame in the roof so that you can look through the rest without fear of crushing her. Just remember what position the frame was in and return it to the same position, facing the same way.

Inspection techniques

You will probably be a little nervous during your first inspection. This is only natural. You don't know your bees very well and they don't know you. You're not yet comfortable handling frames and there is a lot to take in and remember. Hopefully, you are not being bothered by the bees flying around you. You can tell by the sound coming from the hive if they are becoming agitated – angry bees have

tip

If the bees are becoming agitated, cover the hive with cloth and move away from the hive for a minute or two.

a menacing buzz. If you hear this, cover the brood with the cloth and walk away from the hive for a minute or two. Give yourself a generous smoking from the smoker to mask any smells that may be upsetting the bees and return to the hive. A few puffs of smoke into the air above the hive might also help.

Keep a note of what you have seen on each frame. Taking photos enables you to study them at your leisure and find details you may have missed during the inspection itself.

If your colony has been busy and one side of the first and last frames has been drawn out, turn these frames around so that the empty side is facing into the box. This encourages the bees to draw out all of the available space. Do this only if there is no brood on these frames.

Finishing off

Once you have checked all the frames, gently squeeze them back into their original positions. Return the frames that are sitting in the roof. After scraping off any comb and propolis with the hive tool, place the crown board on top of the brood box, moving it gently from side to side as you put it down to stop any bees on the rim being crushed. If there are many on the rim a puff from the smoker will soon get them off into the hive. (When a bee is crushed it emits a pheromone alerting the colony to danger. A puff of smoke should mask the scent.)

If the weather is good and you have seen drawn-out comb, nectar and pollen, then you can stop feeding the bees, even if there is some sugary solution left in the feeder, and put the roof on. If the colony is looking healthy (i.e. there are lots of larvae, capped

brood and food) and the weather is pleasant, open the entrance by another 2.5–5cm (1–2in) to give them more room to come and go.

Walk quietly away from the hive with your smoker and hive tool. Check that you don't have any bees resting on your veil or suit, then take off your protective clothing. Put out the smoker, which may still be smouldering, by sticking a cork or clump of grass in the spout. Then sit down and write up your notes.

Congratulate yourself on a successful hive inspection.

NOTES

At first it may not be apparent why you should keep records, but you will soon find that your notebook (or books – it is a good idea to keep one for each hive) becomes your most important piece of equipment. Make a note of:

- date and time of visit
- weather conditions
- temperament of the bees, especially if they were not behaving nicely
- details on what you saw on the frames, such as good brood patterns, brood and at what stages, queen cells, amount of nectar and honey stores
- your actions, such as marking the queen, feeding the bees, adding or taking away supers, whether you gave any medication or saw any pests
- required action before the next visit, such as making up new supers and frames or returning in seven days to check queen cells.

Records like these show the performance of the colonies and the information can be used to improve management practices for all your hives.

Above *Once you have checked the frames you will need to gently squeeze them back into place, moving gently from side to side to avoid crushing any bees left on the edges. A puff of smoke should remove any stubborn bees that refuse to move.*

Pollen

Honey bees and the flowers they visit are a great example of the interconnected relationships that occur in nature. Plants need bees to pollinate them and bees need plants for food. If this connection were lost, they would both disappear. Thankfully the bee is very good at her job and there are plenty of plants for her to visit. Each plant species has a different nectar which gives each crop of honey its differing flavour, colour and consistency (see 'The taste of honey').

One of the most fascinating sights for any novice beekeeper is the forager bees returning to the hive, their back legs bearing a colourful load of pollen. It looks far too heavy for them to carry, but is actually much lighter than a stomach full of nectar. Although bees are only one of many insects that are responsible for pollination – others include butterflies, moths and even wasps – they are the only ones whose bodies are specially adapted to collect pollen, and the only insects who feed their young pollen, which is rich in the protein, fats and lipids needed for the bees' development.

Pollen is the male gamete of the plant and, in order to reproduce, it needs to mate with a female gamete of the same species. The bee transports the pollen grain from the male to the female by collecting minute particles on the hairs of her body as she visits the flowers, collecting the nectar or the pollen.

It is one of the mysteries of bee communication that a forager bee knows the purpose of her search before she leaves the hive. If she is going out to collect pollen she will take enough honey in her stomach to act as fuel for her flight, plus a bit more to use as glue to bind the pollen together. If she is going out for nectar she can stay out longer, since she will collect the energy-giving liquid en route.

As the honey bee is a methodical pollinator, she generally forages on one type of flower until she has exhausted the supply – unlike butterflies, which can flit from one flower type to another in the course of one flight. Both pollen loads on a bee's legs will therefore be the same colour and from the same source. Back at the hive this doesn't matter, since the pollen is mixed together as food for the bees, but the plants benefit, as this approach gives them the greatest chance of being pollinated.

It is fun to try to guess, with the help of a handy pocket pollen-colour guide, which flower a bee has visited, according to the colour of the pollen and the time of year. Each area has its own pollen cycle, running from early spring through to late autumn. In Britain,

Above *A bee with back legs laden with pollen.*

Right *Fruit growers will hire local bee hives for a few weeks each year to pollinate the crop. Bees will naturally go to the nearest and most bountiful source of nectar, so pollination is easy to promote, as is honey from a single type of flower such as orange blossom, lavender and so forth.*

it tends to start with crocuses, followed by dandelions, fruit blossom trees and horse chestnut, then lime trees, clovers and late-summer plants, before autumn garden flowers such as Michaelmas daisies, and finally ivy. While most pollen is a shade of yellow or orange, ranging from the very pale to the very bright, some is surprisingly brick red (horse chestnut), black or deep purple (oriental poppy) or light green (hawthorn).

A large amount of one particular kind of pollen, however, does not mean that honey will be made from the nectar of that plant. Some plants are a much richer source of pollen than of nectar and vice versa.

When the bee returns to the hive she passes her load to a waiting house bee, who inspects a number of cells before dropping the load into a suitable one. This is usually located just above or beside the brood, so that the pollen is readily available to feed the older larvae, the nurse (newly born) bees and the younger hive bees. House bees will add some honey to the dry pollen and seal the cell with some more. This mixture is called bee bread and it preserves the pollen for later use.

A colony's intake of pollen can fluctuate according to the weather and the availability of pollen-rich plants. To prevent demand outstripping supply during a spell of bad weather, a colony will stockpile around a week's worth of pollen, some 1kg (2¼lb).

tip

Each area has its own pollen cycle, running from early spring through to late autumn.

Commercial pollination

As honey bees can be moved from one place to another, they play a vital role for commercial fruit growers and farmers, and are valued more highly as pollinators than as honey producers. In the USA, for example, where honey bees are moved huge distances to pollinate vast stretches of almond trees in California, apple trees in Pennsylvania and blueberries in Massachusetts, their value to the economy is measured in tens of billions of dollars. In the UK, fruit growers in Kent, jam manufacturers in Cambridge and West Country cider producers will often have their own hives or bring in local apiarists for pollination. As a result, the bees' value to the UK economy is close to £1bn. Lack of bee pollination can result in misshapen fruits and 30 per cent lower yields of many fruits and vegetables. If you have a vegetable patch, you will be pleased to know that bees are particularly good for pollinating marrows, pumpkins, cucumbers and tomatoes.

POLLEN FACTS

- 32kg (71lb) pollen has been recorded from a single colony.
- To rear one worker bee requires ten average bee loads, the equivalent of 0.14g (¹⁄₂₀₀oz) of pollen.
- One average bee load weighs 0.014g (¹⁄₂,₀₀₀oz).
- 450g (1lb) of pollen rears 3,250 bees. This means that 27kg (61lb) of pollen is required to sustain an average colony throughout the year.
- This equates to some two million pollen-collecting trips.

Adding supers

Like anyone who has just moved house, the bees have lots to do to get their new home sorted. During the first couple of weeks, the house bees will be busy drawing out the new foundation, creating the space needed for the eggs, nectar and pollen. The foraging bees will have sought out the best sources of local food and the queen will be laying thousands of eggs. It will soon be time to give the new home an extension, by adding a super. This will give the colony room 'upstairs' to store the honey, freeing up the cells in the brood box for the queen to lay more eggs.

When to add the super

As bees don't work to dates and times, it is impossible to say that exactly two weeks after their arrival your bees will need a super, but what is certain is that you must add a super before the brood box runs out of space. The time this takes can vary from year to year and from location to location, and depends on factors such as the weather, the nectar flow and the character of the bees.

Since there is no rule of thumb as to when to add the super, you need to carry out inspections to monitor the colony's growth every seven days or so. But not more often than that – remember that bees don't like their home being disturbed.

It is unlikely that you will need to add a super after the first inspection as the bees will probably not have had time to draw out the outer frames in the brood box. With each inspection your confidence will grow and you'll start to recognize the different characteristics of each frame and how they change with time. If the weather is favourable your colony will be expanding quickly, so you may be surprised at how crowded and busy the brood box looks when you take off the crown board. Have a super, frames and foundation made up ready for the forthcoming expansion.

You are looking to see if the outermost frames in the brood box have been drawn out. If they have, the colony will soon need more space. If in doubt, add the super anyway: it is better to put it on too early than too late, since lack of super room means the bees will store honey in the brood box. This will make them feel as if they are living in overcrowded conditions and trigger their instinct for swarming (see 'Swarming and increasing an apiary'). You don't want this to happen with your first hive in your first year.

How to add a super

Adding a super is quick and doesn't disturb the bees too much, but you will still need to light your smoker, wear your protective clothing and make sure your hive tool is in your pocket. You'll also need a queen excluder and the super full of newly made frames.

What you are going to do is lift the roof off the hive, take off the crown board, place the queen excluder on the brood box, add the super on top and close the hive again.

Approaching the hive from the side, give the entrance a few puffs of smoke. Wait five minutes for the alarm bells to ring inside, then lift off the roof and place it upside down next to the hive. With your hive tool prise off the cover board (checking the underside for the queen) and place it at the entrance to the hive so that the hangers-on can find their way.

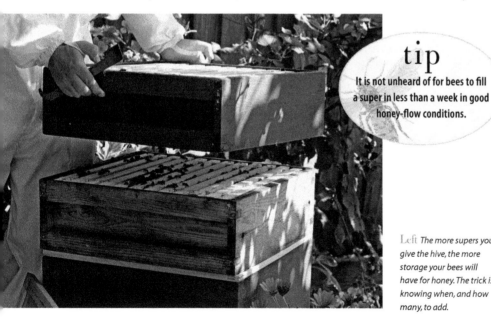

tip
It is not unheard of for bees to fill a super in less than a week in good honey-flow conditions.

Left *The more supers you give the hive, the more storage your bees will have for honey. The trick is knowing when, and how many, to add.*

Right *The queen excluder stops the queen getting into the supers and laying her eggs in the honey stores.*

If there are lots of bees walking around the tops of the brood frames you can give them a puff of smoke to shoo them away. Use your hive tool to scrape away any burr (extraneous) comb that has accumulated on the top of the brood frames. This will make it easier to lay the queen excluder flat on top of the brood box. Gently put on the queen excluder. To keep the bees from getting squashed shuffle it from side to side as you lay it down.

Now pick up the super and place it on top of the queen excluder with the frames aligned the same way as the brood frames. Then replace the crown board and the roof.

It really is that simple and takes a fraction of the time of an inspection.

Adding more supers

One of the skills of beekeeping is knowing when to add another super. Basically, when the first super is more than half full of honey, it is a good idea to add a second. It is not unheard of for bees to fill a super in less than a week in good honey-flow conditions – in early summer when lots of flowers are out, there is no rain or cold spells and there is plenty of sunshine but it is not baking hot. You can add two supers at once if you think one will be full in a week, or if for some reason you aren't going to be able to add another one in seven days' time. If you get your bees early in the season and the honey flow is good, you may have to add four supers to your hive over the course of the summer.

Supers can be added by either placing each new one on top of the last (top supering) or by placing the new one between the brood box and existing supers (bottom supering).

Top supering is popular because it requires less work, while bottom supering may encourage more honey stores as the bees are closer to the empty frames. There are no strict rules and no best way of doing it – it is another example of the art of beekeeping.

Bees more readily accept honey supers with drawn-out foundation than those containing frames of new foundation, because they can immediately store nectar in the drawn-out cells. In your first year of beekeeping you will not have any drawn-out foundation (this comes after you have spun off the honey). So your first supers will contain clean foundation. However, after harvesting your first super (see 'How to harvest honey') you will be left with frames full of drawn-out foundation which you can use in future supers.

> **Adding a super gives the colony more room 'upstairs' to store the honey, freeing up the cells in the brood box for the queen to lay more eggs.**

Sometimes the bees are hesitant about moving through a queen excluder, so if you find on your next inspection that they haven't begun to draw out the foundation above it, take it off for a week to give them the freedom to move into the super unhindered. Once the bees are working the super, replace the queen excluder and you should find that they will pass through it now they have the smell of the new cells to encourage them. Soon your super will be filled with the lovely sight of white-capped honey stores. You can keep these supers on the hive until you are ready to harvest them.

Right *By the end of the summer your hive could be almost 6ft tall with all of the supers you have added.*

Gardening for bees

Nothing evokes summer quite as much as the buzz of bees collecting pollen. Unfortunately, some of our most popular cultivated flowers provide no nectar and little pollen. Encouraging honey bees to stay in the garden or attracting other types of bees may require some horticultural changes. Culinary herbs and fruit trees are bee-friendly, and many wildflowers often regarded as troublesome weeds provide a rich source of food. Leaving a patch of your garden to run wild can bring much pleasure to your bees and be a welcome habitat for other creatures. The most useful flowers you can plant are those that provide food early and late in the year. Not only will they supply nectar and pollen to your bees when these are in short supply, but they will also add colour to your garden.

Gardens and flowers

You may want to give your bees a helping hand when it comes to collecting pollen and nectar by planting bee-friendly flowers and shrubs in your garden or outside space. It is unlikely that you will have enough flowers, even in a larger garden, to produce sufficient nectar to sustain a hive, so your bees will probably go further afield – up to 5km (3 miles) – to find richer and denser sources of food. But the right flowers, fruit and vegetables in your garden will attract some of your own bees, as well as welcoming other visitors such as the bumblebee and solitary bee species, especially if you have made homes for them (see 'Bee species'). Above all, a garden or patch devoted to plants that are attractive to bees can be a source of great pleasure to any beekeeper, as much for the riot of colour as for the activity of the bees.

Two other important factors contribute to a successful bee garden: the flowers should be in full sunlight and they should be planted in groups. Flowers grown singly or in twos and threes may fail to attract bees. A decent-sized clump of a suitable plant such as lavender is much more valuable. Likewise, bees often overlook flowers grown in shade, even though they produce nectar and pollen.

Don't place the plants too close to the hive thinking that you are helping the bees by reducing the distance they have to travel to find food. Flowers very near a hive are in the firing line when the bees go on their toilet trips, so they sensibly avoid them as a food source.

Unfortunately, some of the most spectacular-looking garden flowers are of no use whatsoever to the honey bee. Double-headed roses, chrysanthemums and dahlias, for example, provide no nectar and hardly any pollen. In contrast, many flowers that are often discounted as weeds, such as dandelions and forget-me-nots, provide a rich source of food. That is why one of the best and easiest things you can do to make your garden more bee-friendly is to throw away the weedkillers that maintain those immaculate-looking lawns, and let your lawn and flowerbeds go wild.

If you are not quite ready to hand over your well-tended garden to the vagaries of nature, the next best thing is to leave a patch to run wild. One way to get your wild garden started is to sow wildflower seed mixtures. The flowers will be a useful source of nectar and pollen, but your wild garden will not just attract bees – the tall grass will also provide a welcome habitat for a number of other creatures and

Above Leaving a patch of your garden to grow wild with plants such as poppies, daisies, dandelions and cornflowers will look lovely and provide good sources of food for bees.

Left Lavender is hardy, easy to grow, smells lovely and above all is rich with food for bees.

the berries, fruits and seeds produced by some of the plants will be food for birds and other animals. Soon this part of your garden will be teeming with life.

In addition to wildflowers, the most useful plants you can grow are those that provide nectar or pollen very early in the year when it is in short supply in the hive and elsewhere, and those that flower late in the autumn, providing food for the winter. Bees thrive when there is a continuous supply of nectar and pollen throughout the brood-rearing season.

Global warming is thought to be causing a change in flowering times, with plants flowering earlier and some flowering more frequently throughout the winter months in temperate climates. The potential impact on beekeeping is that the bees' natural rhythm of seasonal population decline will be confused by the availability of nectar at the time of year when the queen doesn't normally lay eggs. This availability of nectar will encourage her to

tip

The potential impact of global warming is that the bees' natural rhythm of seasonal population decline will be confused.

lay, but there won't be enough workers to look after the brood, which will lead to weakened and stressed colonies. It will take generations of bees to evolve to fit in with the new climatic conditions. And a worrying impact of hotter and drier summers is a likely change in garden styles. People swapping their poppies for palm trees or their sweet peas for succulents could have a significant effect on the availability and timing of bee forage, not to mention the increasing trend for easy-maintenance decking and paving replacing labour-intensive and water-hungry lawns and flowerbeds.

Year-round food

Snowdrops, the buttercup-like winter aconite, crocuses, daffodils and hyacinths all supply much-needed fresh pollen and nectar after the long winter months. These can be planted under trees or shrubs to produce a carpet of flowers in early spring when large parts of the garden are still bare. The white deadnettle, which flowers from this time well into the winter, and increasingly all year round, is one of the bee's best friends, as are other year-round plants such as ivy and heathers.

Purple- and white-coloured buddleia, Michaelmas daisies and sedum, which turns red come the autumn, are all late-flowering plants which, as well as supplying the bees' final sustenance before winter, add a late flourish of colour to your garden.

Flowers to consider from late spring through the summer include tulips, forget-me-nots and dandelions, which are out just before the fruit blossoms appear. A bed containing the purple spikes of *Salvia x superba* and edged with the long-flowering common catmint is sure to be full of bees in midsummer. Other perennials you could usefully plant include fuchsia, cornflowers, yarrow, golden rod, geranium, the bell-like campanula and the fast-growing lavatera with its pink, trumpet-shaped flowers.

If space allows, hazels, shrubby willows and honeysuckles are useful sources of nectar and pollen in early spring. In early summer, a wall or fence covered in red-berried cotoneaster will be a favourite. Among taller plants, hollyhocks, sunflowers, foxgloves and the daisy-like rudbeckia all attract bees.

In a wilder setting, clovers, rosebay willowherb and brambles are among the midsummer pollen producers. For autumn pollen, try heathers, thistles, ivy, balsam and autumn-flowering crocuses.

Many culinary herbs are a good source of nectar, so as well as creating a herb garden for your kitchen you will simultaneously create a bee garden. The mint family, which includes sage, thyme, marjoram and basil, is a must, along with rosemary and lavender. Herbs grow well in pots, so if you have just a backyard, a roof terrace with no space for flowerbeds or even just a windowsill, you can still attract bees with a collection of pots planted with a variety of herbs. You could also add a miniature fruit bush – these too are now available in pots. Raspberry, blackberry and gooseberry bushes will provide not only juicy soft fruits for you to eat but also delicious nectar for the bees. In the case of raspberries, bees are known to collect the sweet juice from the ripe and overripe fruit because unlike other fruits, they can puncture its delicate skin. Apples, for example, are of no interest to a bee because the skin is too tough, but the delicate white blossom on the apple trees in spring is a welcome source of food early in the season.

Other plants that are easily grown in a small outside space, or in pots, and provide nourishment to both beekeeper and bees include beans, marrows and tomatoes. And if you grow sweet peas intertwined with the beans, in true cottage-garden style, the bees will have two flowers to feast on when they visit.

You will often hear beekeepers talking about a nectar gap. This refers to the short period in the summer when supplies of nectar from fruit trees, such as apples, and oil-seed rape have dwindled and other sources are not yet available. Bear this in mind when you are planning what to plant in your bee garden.

Trees

If you have diligently planted a variety of bee-friendly flowers, shrubs and fruit bushes, you can take great delight in frequent visits from honey and bumblebees throughout the year. But if you live in a town or a suburb, the chances are that much of your honey will actually come from nectar collected from surrounding trees. In the countryside where there are fields of crops and pastures of wildflowers, trees are less important for bees, but in towns and cities they will probably provide their main source of food. Look around at the trees that shade your garden, the trees in your streets and those that fill nearby parks. Even in the inner cities you'll be surprised just how many of them dot the landscape. And tree-lined avenues give many suburbs their 'leafy' prefix.

Hazels, alders and poplars supply pollen at the beginning of the year. Acers, including the ubiquitous sycamores, follow with both pollen and nectar. At the same time there is blossom from the horse chestnut and sweet chestnut trees. For many town dwellers, however, it is the lime tree which flowers in midsummer that is the source of much honey.

> Many culinary herbs are a good source of nectar, so as well as creating a herb garden for your kitchen you will simultaneously create a bee garden.

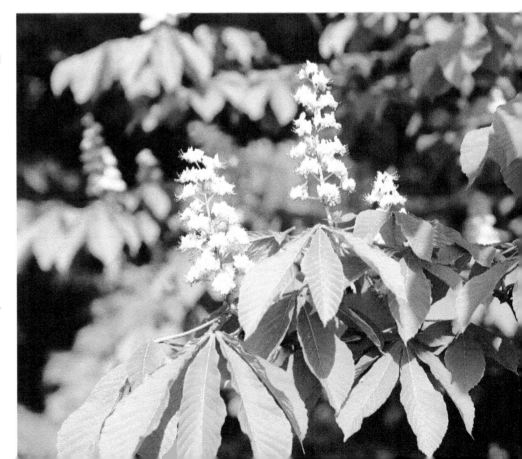

Some suggested flowers for a bee garden

Annuals

Baby blue-eyes (*Nemophilia spp*), bindweed (*Convolvulus* spp.), blanket flower (*Gaillardia* spp.), blue-eyed Mary (*Collinsia violacea*), California poppy (*Eschscholzia californica*), candytuft (*Iberis amara*), China aster (*Callistephus chinensis*), clarkia (*Clarkia elegans*), common baby's-breath (*Gypsophila paniculata*), cornflower (*Centaurea cyanus*), flax (*Linum* spp.), gilia (*Gilia* spp.), godetia (*Clarkia amoena*), lavatera (*Lavatera* spp.), love in the mist (*Nigella* spp.), mallow (*Malva* spp.), malope (*Malope trifida*), meadowfoam or poached egg plant (*Limnanthes* spp.), Mexican aster (*Cosmos bipinnatus*), mignonette (*Reseda* spp.), Nasturtium (*Tropaeolum* spp.), phacelia (*Phacelia* spp.), pheasant's eye (*Adonis* spp.), sweet alyssum (*Lobularia maritime*), tickseed (*Coreopsis* spp.).

coneflower (*Rudbeckia* spp.), cranesbill (*Geranium pratense*), fernleaf yarrow (*Achillea filipendulina*), fleabane (*Erigeron* spp.), forget-me-nots (*Myosotis* spp.), French honeysuckle (*Hedysarum coronarium*), fuchsia (*Fuchsia* spp.), golden rod (*Solidago* spp.), herb hyssop (*Hyssopus officinalis*), hollyhock (*Alcea rosea*), horehound (*Marrubium vulgare*), hound's tongue (*Cynoglossum* spp.), knapweed (*Centaurea* spp.), large blue alkanet (*Anchusa azurea*), lavatera (*Lavatera* spp.), lavenders (*Lavandula* spp.), loosestrife (*Lythrum* spp.), marjoram (*Origanum majorana*), sage (*Salvia* spp., especially *S. x superba*), scabious (*Scabiosa* spp.), sneezeweed (*Helenium* spp.), thoroughwort (*Eupatorium* spp.), thrift (*Armeria maritima*), thyme (*Thymus* spp.), veronica (*Veronica* spp.), willowherb (*Epilobium* spp.), woundwort (*Stachys* spp.).

Perennials

Alpine rock-cress (*Arabis alpina*), alyssum (*Alyssum* spp.), aubrieta (*Aubrieta* spp.), Canterbury bells (*Campanula medium*), Carpathian harebell (*Campanula carpatica*), catmint (*Nepeta* spp.), checkerbloom (*Sidalcea* spp.),

Bulbs

Autumn crocus (*Colchicum autumnale*), common hyacinth (*Hyacinthus orientalis*), crown imperial (*Fritillaria imperialis*), quamash (*Camassia* spp.), snowdrop (*Galanthus nivalis*), snowflake (*Leucojum* spp.), tulip (*Tulipa* spp.), winter aconite (*Eranthis hyemalis*).

tip
The most useful flowers you can plant are those that provide food early and late in the year.

Right *It is most beneficial for the bees if you plant a variety of flowers that between them will provide year-round food.*

All about honey

It is probably the prospect of home-grown honey that entices most people to keep bees. The sweet golden liquid that has enchanted mankind for centuries is one of nature's glorious gifts. What could make a more personal present than honey from your own hive? One of the most fun times for all the family is the honey harvest, when the oozing substance is ready to eat. When you will get most of your honey, how much you will get, and what it will taste of are often difficult to predict. Your bees will constantly surprise you. How you harvest and jar your honey will depend on whether you intend to keep it, sell it or show it. Whatever you decide, the right preparation will make all the difference.

How bees make honey

Honey starts as nectar – a watery sugar liquid mainly found in the flowering parts of pollinating plants. This attracts nectar-loving animals and on their visits to drink the liquid they inadvertently collect and disperse the plant's pollen. This in turn ensures the flower's reproductive success, while giving the pollinators a high-carbohydrate energy drink. Hummingbirds, butterflies and ants also love nectar, but the honey bee does more than just drink it – it makes it into honey and stores it for future use.

A worker honey bee can carry the equivalent of half her own body weight of nectar back to the hive in her honey stomach, and during that journey she adds enzymes which begin changing the nectar's chemical properties from a complex to a simple sugar. On returning to the hive she regurgitates the nectar and passes it to a house bee, which continues to break down the sugar by placing a drop on her proboscis and 'chewing' while exposing it to the air. She may 'chew' on one droplet of nectar for 20 minutes. This process gets the enzymes working and reduces the water content of the nectar.

The bee then hangs the droplet inside a cell and repeats the process with another, continuing until the cell is full. To reduce the water content further, the house bees circulate air around the honeycombs, using their wings as fans. The air sucks the water out of the liquid, similar to a chef reducing the liquid content of a sauce to thicken it. This process continues until the water content has been reduced to around 19 per cent, creating the thick, sweet substance we call honey.

Once the bees are happy that the water content is correct, they cap the cell with a thin layer of fresh wax, preserving the honey for future use. A frame of freshly capped honeycomb is a glorious sight and highly prized. You will sometimes find whole frames for sale and at honey shows they are judged on their purity and quality.

Above *Honey is the bees' energy source, and throughout the summer they collect enough to feed the colony over the cold, winter months.*

Above Right *Bees fan their wings to cause a draught in order to ventilate the hive, and dry any evaporating moisture from the nectar.*

Right *The bees store nectar in the hexagonal cells of the honeycomb. They reduce the water content to around 19 per cent to create the substance we call honey.*

HONEY-MAKING FACTS

Bees are very fuel-efficient and can fly 1km (⅝ mile) on 0.5mg (0.000017oz) of honey. A foraging flight may last 45 minutes and in that time a bee can visit 100 flower heads, which will give her 50mg (0.0017oz) of nectar – about half her own body weight. If she managed ten trips a day she would bring 0.5g (0.017oz) of nectar back to the hive. If each of the 10,000 foraging bees did the same, some 10 million flower heads would be visited in one day (that doesn't mean 10 million flowers, of course – the same flower would be visited more than once) and the bees would return to the hive with 5kg (11lb) of nectar. After processing, this would make 1.25kg (2¾lb) of honey. This doesn't happen every day but, when conditions are conducive to collecting nectar, bees really are this busy.

Right *To ensure that the honey is fresh for the winter store, the bees make a white wax cap over it, which we can see here.*

Throughout the summer, the bees collect enough nectar to store and feed the colony during the cold, nectar-drought months of winter. It is these stores that you are in effect stealing from the hive when you harvest the honey.

A strong colony of honey bees gets busy collecting nectar as soon as the weather is warm enough for the plants to start producing it. When the fruit trees have blossomed and the dandelions are out you can be sure the foragers will be out and about too. They may travel up to 5km (3 miles) from the hive in their search, but are thought to favour the nearest source. When a scouting foraging bee finds a good source of nectar, she uses the waggle dance to let the other foragers know (see box). Bees are not fortune-tellers and the vagaries of the weather can produce periods where collection is not possible – a rainy spell or even a heat wave, when flowers conserve their nectar – so they collect as much as they can, when they can. The amount depends on the availability of the nectar and how much storage space is in the hive.

As well as being used to feed the larvae, honey gives the foragers fuel for their nectar- and pollen-collecting flights, and the house bees energy to keep the brood areas of the hive warm. To keep the maternity wing and nursery at a constant temperature of 35°C (95°F) the bees huddle together and shiver. This movement of the body, which increases the temperature to the required heat, takes a lot of energy.

tip

Bees are not fortune-tellers and the vagaries of the weather can produce periods where collection is not possible.

If the outside temperature is too high, the bees use their wings to fan air throughout the hive to keep it cool. Again this uses a lot of energy and they get this by eating the honey they have been so busy collecting and storing.

Bees also need to eat honey to stimulate the production of wax.

During a year a colony requires at least 120kg (265lb) of honey just to be able to function.

THE WAGGLE DANCE

The returning forager bee is laden with her bounty and eager to tell her sisters where the supply can be found. First she climbs on to the vertical comb and lets the other bees taste a bit of the treasure – this could be nectar, water, pollen or propolis. Once they have had a good sniff to assess the quality and know what they are going to be looking for, they stand around watching, listening and smelling the bounty hunter as she does her waggle dance.

This shaking of the body as she repeats a 'figure of eight' pattern tells the onlookers the direction and distance of the new supply. If the bee waggles her body for a fraction of a second the source is near, but if the waggle lasts for a couple of seconds the source will be around 2,000m (1 ¼ miles) away. The direction of the source is conveyed by the direction of the dance in relation to the sun's position. Think of a clock face on the honeycomb. If the dancer's head points upwards towards 12 o'clock, she is telling the other foragers that the source lies on a line directly towards the sun. When, after an hour, the sun has moved 15 degrees across the sky, the forager will need to alter her dance direction by 15 degrees. So this time the dance will not be straight up on the honeycomb; instead, the head will be pointing halfway between 11 and 12 o'clock. When the new foragers fly from the hive they first get a sense of where the sun is and then fly in the direction told to them by the waggle dancer – in this case 15 degrees to the left of the sun. If at another time the source of food was opposite the sun, the dancer would waggle with her head pointing downwards.

The waggle dance isn't totally accurate, but enough to get the bees near the new source. When they get there they may have to hunt around for a few minutes before they recognize the food by its smell.

It is a remarkable form of communication for any animal, let alone one with a brain the size of a breadcrumb. We mustn't forget to thank Karl von Frisch, who won a Nobel Prize in 1973 for his work on this subject.

The waggle dance isn't totally accurate, but enough to get the bees near the new source.

How much honey?

During the course of the summer the colony will have been expanding as the queen lays thousands of eggs to increase the army of foragers who bring back the food to feed the hive. In the wild, the honey the bees produce at the beginning of the season is used by the emerging population, which increases five-fold over the summer. Towards the latter half of the year the queen reduces her laying and the colony prepares for winter, getting the stores in to keep them going over the cold months.

Things are a bit different for a colony that is kept in a hive and managed by a beekeeper. Since you can keep adding extensions to the hive, the bees find themselves with more room to fill. It is their instinct to fill available space with honey stores and the more space available the more eggs the queen will be encouraged to lay. So the amount of honey you get from a colony depends in part on how you manage the hive. Other factors that play a role include the weather, the availability of good forage, competition from other hives in the vicinity, any ailments the bees may have and the character of the bees themselves.

It is estimated that you can expect anything between 9 and 18kg (20 and 40lb) of honey from one hive in your first year of beekeeping. Just one National super with ten frames of capped honey can weigh around 10kg (22lb). So even if you think you have bought plenty of jars, don't be surprised if you end up having to borrow containers of all shapes and sizes from family and friends.

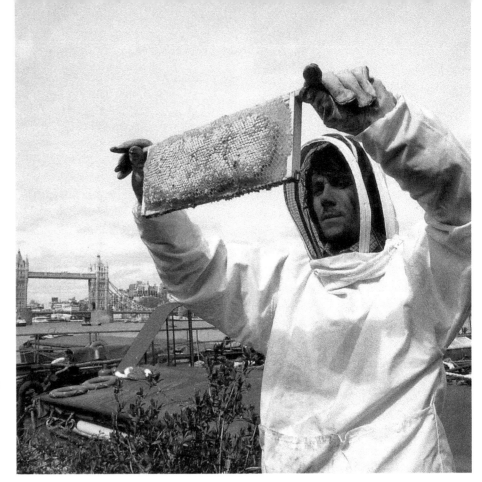

Above Cities are awash with nectar for the bees to feast on. With a huge diversity of flowers and plants to visit, urban honey is often more flavoursome than other varieties.

Right Oil-seed rape covers huge swathes of land with its bright yellow flowers which bees love, quickly making honey from it in late spring.

When you get the honey

When your honey is ready to harvest depends primarily on the types and abundance of flowers, crops and trees within a 5km (3 mile) radius of your hives. But so many other factors come into play – most importantly the weather – that it is impossible to promise anyone a jar of home-grown honey by a certain date. If the colony is strong and the conditions have been good, you may find that a super or two will be full by mid summer. You may also get a late harvest towards the end of the summer. Each hive will perform differently.

Even two hives standing next to each other can produce different amounts of honey at different times.

It is also hard to predict what your honey will taste of unless the hives are located near fields of a particular source of nectar. If, for example, there is oil-seed rape nearby, there is little chance of your honey tasting of anything else. Bees love this bright yellow flower which is increasingly taking over vast swathes of the countryside. Its nectar is available in the spring and they can work it very quickly, so you may get a super's worth of oil-seed rape honey by late spring. However, the downside of oil-seed rape is that it crystallizes quickly and the bees then find it difficult to eat. Remove it within a few days and extract it (see 'How to harvest honey') before it crystallizes on the frame.

The predominance of oil-seed rape in some parts of the countryside means that many city honeys are more flavoursome and delicate. Urban bees, having enjoyed the biodiversity of plants and flowers that gardens, parks and railway cuttings have to offer, can often produce honey with a much richer texture and taste than their rural sisters. Lime trees

line many city streets, so it is no surprise that late-summer honey from an urban hive often has a slight citrus taste. A bumper harvest of elderflower honey in early summer after a very wet spring is more unusual. Bees turn to elderflower only in desperation, but the honey is delicious with a light, fragrant bouquet – the one consolation for all the rain!

Not knowing what your honey will taste of from one harvest to the next, let alone one year to the next, makes the honey harvest really exciting as you try to guess what plants your bees have been feeding on and how the

honey will compare to previous batches. The downside is that you can't promise people more of, say, the early summer crop they particularly liked the previous year, because you can't guarantee that this year's honey will taste anything like it.

Unlike wine tasting, there don't seem to be courses in honey-tasting around, but you can still enjoy the challenge of trying to decipher the nuances and subtle flavours of your polyflowered honey.

The taste of honey

Honey's subtle taste and flavour is mainly determined by the plant from which the nectar is obtained. This also shapes the aroma, density and colour of the honey. In many countries, honey comes from a single plant such as clover in Canada and New Zealand, eucalyptus in Australia, leatherwood in Tasmania, lavender and rosemary in France or Spain, wild thyme in Greece, buckwheat in the eastern US and Russia, orange blossom in Florida or California, and acacia in eastern Europe. In Britain, most honey – with the exception of heather and some white clover honey – is usually a blend of wild flowers, including hawthorn, fruit blossom and the clover-like sainfoin. However, in urban areas, where trees in parks and lining roads are the predominate source of nectar, slight citrus-flavoured honey from the lime and the sycamore tree are common. There is no direct correlation between the flavour of a honey and its colour, but the lighter the honey, generally the milder in flavour it is than darker-coloured varieties.

Acacia tree	Very light yellow, almost transparent
Alfalfa	Amber to beige
Apple blossom	Light amber
Balsam	Amber
Borage +	Pale yellow
Buckwheat	Purplish black
Chestnut	Very dark brown
Clover	Varies from white to light amber
Dandelion #	Deep yellow
Elderflower -	Very pale yellow
Eucalyptus	Light amber
Fir tree	Dark brown
Lavender	Dark amber
Leatherwood	Amber
Lime blossom *	Very light amber with greenish tinge
Hawthorn	Dark brownish
Heather	Reddish brown
Orange blossom	Light amber
Oil-seed rape #	Light yellow
Raspberry	White with a red hue
Rosemary	Light amber
Sainfoin	Very light, almost white
Sycamore tree *	Amber with greenish tinge
Sunflower #	Opaque, egg-yolk yellow
Thistle	Deep yellow with green hue
Wild thyme	Amber
Willowherb x	Pale, water-white

+ Also known as star flower honey as borage has sky-blue star-shaped flowers.
\# Honey crystallizes very quickly so has to be harvested early.
- Very unusual but in a particularly bad Spring when no other nectar sources around the bees will visit the elderflower.
* One undesirable feature of the lime tree and the sycamore is honeydew, which is the secretion of insects that live on the leaves. On occasions the bees collect the honeydew, but in many European countries, especially Germany, the most highly sought-after honey is honeydew from conifers.
x Also known as fireweed.

Very mild, hint of vanilla	Thin	Mid summer
Delicate and fruity	Thick	Late summer
Hint of apple	Thick	Early summer
Toffee flavoured	Thick	Late summer
Mild, hint of camomile	Thin	Mid summer
Strong, malty	Very thick	Early summer
Treacle and molasses, bitter, rich in pollen grains	Very thick	Late summer
Creamy butterscotch	Medium	Mid summer
Hint of ammonia	Thick	Spring
Elderflower cordial	Thin	Spring
Hint of toffee and raisin	Thick	Summer
Malty, toffee-like, with hint of aniseed	Thick	Summer
Strong 'flowery' perfume	Medium	Mid/late summer
Malty, spicy, woody aftertaste	Medium	Mid summer
Citrus/minty	Very thin	Mid summer
Rich, nutty, almondy	Very thick	Early summer
Hint of toffee	Thick, jelly-like	Late summer
Hint of citrus, tangy and very sweet	Medium	Late spring
Bland	Medium	Late spring
Raspberry	Unctuous	Mid summer
Strong aroma, hint of rosemary	Medium	Early summer
Slight fruity taste	Thin	Mid summer
Hint of citrus	Fair	Mid summer
Suggestive of lemon	Thick	Late summer
Slightly bitter and astringent	Medium	Late summer
Intensely aromatic, spicy, touch of thymol	Medium	Late summer
Bland	Medium	Late summer

How to harvest honey

It has been said that the enjoyment of beekeeping can be spoiled by the hassle of harvesting honey, but with some planning and the right equipment the process can be quick, straightforward and above all lots of fun – as long as you accept the fact that not only your equipment but you too are going to get sticky. Fortunately honey is easy to clean off with warm soapy water. Wax is a bit more troublesome, so have plenty of newspaper or old tablecloths around to put on your worktops and floor, and keep dedicated bowls and utensils for the job.

Ideally you will have a utility room with a sink and lots of space in which to harvest your honey, but we don't all have that luxury. You may well have to work in a small kitchen, in which case keep the doors and windows closed to keep out any unwelcome intruders, particularly inquisitive wasps. Try to avoid harvesting your honey outside, as you will soon be surrounded by bees and wasps.

The plan is to get the honey out of the frames into containers without bits of wax or bees in it, so you need to work in a systematic way: get your frames full of capped honey into the house without the bees, open up the cappings, extract the honey, strain it to get rid of the unwanted bits, then let it settle before pouring it into jars. Remember that full supers can be very heavy – around 10kg (22lb) – so if you are a lone beekeeper you may need some assistance carrying them.

Getting the frames without the bees

The first step is to get the bees out of the super that you want to harvest. You really don't want lots of bees clinging on as you try to remove the honey from the frames.

To do this, you can use a clever little device called a bee escape. There is a variety of types on the market but they all work on the same principle – they let the bees leave the super but don't let them return. In other words, they have a one-way valve.

Place the bee escape in the hole of the crown board and place the crown board under the super you want to remove. After a few hours the bees will have moved downstairs and been unable to get back up, so the super should be bee-free.

Comb honey

Once you have got your super safely inside, it is time to decide if you would like to keep some comb honey, or if you want to extract all the honey from the comb for liquid honey. Most people like a bit of both.

For comb honey, simply use a thin, sharp knife and slice small chunks, perhaps 10 x 7.5cm (4 x 3in) of the comb straight from the freshly collected frames. If you have been using wired foundation, cut the wire and pull it out of the comb before putting the comb into your containers. If you think that you may want more comb honey in the future, you will need to buy unwired foundation. The wires are only there to give strength when the honey is spun in an extractor (see below). You may like to put some comb in a jar and fill it up with liquid honey. This is called chunk honey and it makes a great present for people who have expressed a preference for honey on the comb.

Scraping off the honey

If you don't have access to an extractor, you can harvest your honey by scraping the comb. It takes longer and is messier than using an extractor, and it uses more utensils, but has the slight advantage that less air gets mixed in with the honey. The main disadvantage is that it destroys the honeycomb. Ideally you want to give this back to the bees, as it cuts down their workload: instead of having to draw out new foundation, they are able to store nectar immediately in the honeycomb they made during the year.

If you are going to use the scraping method, you will need a large pot to collect

---

HARVEST AND CHILDREN

You will find that children will be fascinated by your bees and the hives. Any initial caution will soon give way to interest, as they become more comfortable around them and feel safer.

As well as the hive inspections and day-to-day maintenance, the honey harvest is another time when children can easily get involved, and beekeeping can really become a family activity. Naturally the thought of young children being covered in sticky honey might be a bit daunting, but with carefully allocated roles for children and adults, there is no reason why they can't take part happily, safely and cleanly.

Make sure that you are fully familiar with the harvest process, so that you can be confident of the tasks that children can undertake.

Above *Empty honeycomb after the honey has been extracted.*

the honey, a large container for the comb, a large spoon or scraper and a coarse sieve. Later you will need a finer sieve to filter the honey a second time and another large pot or bucket – preferably with a valve (honey-gate) at the bottom so your honey can easily flow into jars. These honey buckets or tanks can be bought from many beekeeping suppliers. Place the coarse sieve over the large pot and, holding the frame of honeycomb over the

sieve, scrape the honey straight off the frame, comb and all, into it. You are not going to be able to use the foundation again, so just scrape away as hard as you like.

The sieve will soon be filled with a lovely gooey pile of comb and honey. Let the honey ooze through the sieve into the holding pot. Once most of it has passed through, empty the wax comb scraping into the large container. (You can sieve this again later and use the remaining wax for making candles or lotions – see 'Bee

products'.) Continue this process until all the frames have been emptied.

To get really clean honey, filter it through either a finer sieve or some muslin into another large holding pot or bucket. This will weed out the small particles of wax, bee bits and pollen that got through the coarse filter.

After the second sieving process is complete, put a lid on the container and let the honey stand so that the air bubbles can rise to the top. This may take as much as a couple of

tip
You can sieve the wax comb scraping again later and use the remaining wax for making candles or lotions.

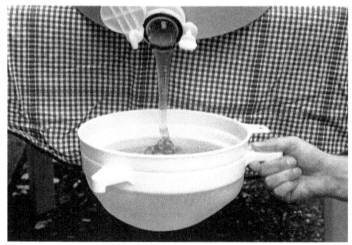

days, depending on the consistency of the honey and the ambient temperature.

You now have empty scraped frames that are wet with honey. The best thing you can do with these is to return them to the bees for cleaning. Put them back in the super and place the super above the crown board (as long as the hole is open). In a day or two the frames will have been licked clean and you can use them again to put new foundation in.

Using an extractor

An extractor makes the process much easier, quicker and less messy. It can be costly to buy, so it is worth asking your local beekeepers' association if they have a communal extractor for hire. Many do, but there may be a waiting list at weekends at busy times of year.

There are a variety of extractors on the market. The cheaper ones are manual and plastic. More expensive versions have an electric motor and are stainless steel. In either case, extractors are either tangential or radial. In a tangential extractor you have to turn the frames in order to empty the honey from both sides. Radial extractors do both sides at once. If you have no more than a handful of hives, a manual tangential extractor will suffice.

Whichever one you decide to use the principle is the same. A number of frames of uncapped comb are placed in the device and spun either by you or by the motor. The honey 'flies' out of the comb on to the sides of the extractor, collecting at the bottom of the tank. The collected honey can then be filtered into another honey tank or bucket for settling and later bottling.

For this method of honey extraction you will need an uncapping device (either a knife or a fork), a honey tank/bucket – preferably with a honey-gate at the bottom, as this makes it easier to pour the honey into jars –

and a double strainer. The uncapping device opens the caps of the honeycomb, allowing the honey to spill out of the frames when spun in the extractor. It's up to you whether you choose a fork or a knife – try both and see which you are more comfortable with.

Run the fork or knife just under the surface of the cappings. You want to take off just the white cappings (left) and leave the rest of the comb intact. An uncapping knife works better when it is hot, so have some water on the boil to dip it into after each scraping.

Put the wax cappings into a container – you can melt them later to make candles. When the cappings have been removed from each side of the frame, place the frame into the extractor.

Extractors normally hold at least four frames, so continue uncapping until the extractor is full. Put on the lid and turn the handle (or flick the switch) slowly at first. Let some of the honey spin out before increasing

the speed (so that the comb stays intact) and continue spinning for a couple of minutes. With a tangential extractor now is the time to rotate the frames to empty the honey from the other side, then carry on spinning for a few more minutes.

You will see that the honey is clinging to the side of the tank and collecting at the bottom along with bits and pieces of wax and pollen. When no more honey is coming off the frames and the spinning has become lighter, remove the comb and store it in the super (you can give it back to the bees to clean).

When all the frames have been spun, filter the honey through the double sieve into the honey tank or bucket (or any large container). It will take around half an hour to sieve 9kg (20lb) of honey – about one full super's worth. Now cover and leave to settle for 24 hours or more to let the bubbles rise to the surface. Once they have been skimmed off your honey is ready to pour into jars.

Right *When all of the frames have been spun, it is important to filter the honey through a double sieve into a honey tank or bucket. This will remove any bits in the honey, such as wax or bee parts. It will take around half an hour to sieve 9kg (20lbs) of honey.*

How to jar honey

Choosing your jars

Your choice of jars depends largely on what you plan to do with the honey: are you going to eat it yourself or give some away as presents to family and friends; are you going to show it, or even try to sell it? If it's mainly for your own use, you may just want to store it in recycled jars that you have been keeping hold of throughout the year. A note of caution when using old jars, though: however much you wash them, lids often retain the smell of their previous contents and this will ruin the taste of the honey.

If you want friends and family to be able to share the enjoyment, you may want to buy a number of identical small attractive jars. Searching the internet should reveal suitable suppliers who will deliver to your door. Look for ones with tin-plate lids with a flowed-in seal (a built-in annular ring on the underside of the cap) which prevents air and moisture getting into the jars and ruining the honey.

Jars like this are also perfect if you want to sell your honey at a local delicatessen or farmers' market. If you want to show your honey, the type and size of jar will be stipulated in the entrance rules of the show (see right).

Putting your honey in jars

A mistake beginners often make in their excitement is not leaving the honey to settle for 24 hours or more in a honey bucket or tank. If you pour it into jars immediately after filtering, it will contain lots of bubbles. While these don't seem to affect the flavour, they do float to the top and leave an unsightly thin foam on the surface.

Once the honey has settled, pour it into the jars quickly – to reduce the amount of exposure to the air – but not so fast that air gets trapped. Getting this right becomes easier with practice and is a matter of judging how wide to open the honey-gate at the bottom of your honey bucket or tank. If your container does not have a valve, transfer the honey into a jug and then pour it from the jug into jars.

After you have filled a jar, put the lid on immediately. When they are all full store the jars in a dark place, such as a cupboard or larder, at room temperature.

Labelling your honey

It's a good idea to label jars with the location of the hive, and the month and year the honey was made. If you are selling your honey or giving it away, include your contact details in case someone wants another jar. You can design your own labels on a home computer and print them out on sticky labels. It's not difficult to make them attractive as well as informative.

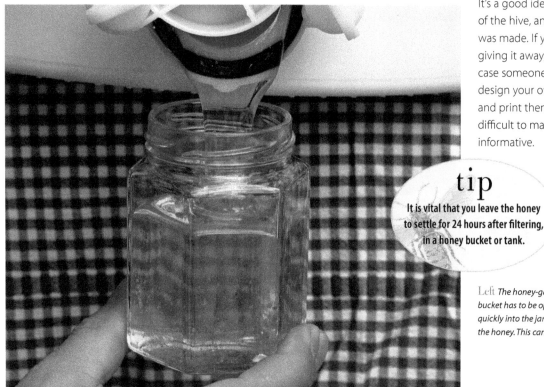

tip
It is vital that you leave the honey to settle for 24 hours after filtering, in a honey bucket or tank.

Left *The honey-gate, or valve, at the bottom of the honey bucket has to be open wide enough for the liquid to pour quickly into the jars, but not so quickly that air gets trapped in the honey. This can take some practise to get right!*

How to show honey

You may be curious to know just how good your honey is. Even though everyone who has tasted it says it looks delightful and it has a delicious flavour, how does it compare to other honey? Does it really have an exquisite taste or is it run-of-the mill? One way to satisfy your curiosity is to enter a honey show. Ask your local or national beekeepers' association or check the beekeeping journals. Most shows have classes open to beginners and children to encourage newcomers to have a go. Most charge an entrance fee and each class will have its own strict rules, so be sure to read them carefully and follow them to the letter.

Showing honey is not just about trying to win a ribbon, though that will of course be a tremendous accolade; it is also about meeting your peers, getting advice from the experts and learning how to improve your skills in honey production.

To be in with a chance of getting on the podium all aspects of your product must be top class, from the labels, which must conform to the show's guidelines, to the spotlessly clean jars, and the honey will need to be free of any unwanted materials such as bee parts or dust. Honey judges, especially at the top levels, will be able to tell if your honey has been spoiled by the soot of your smoker or scorched by an electric uncapping knife in the harvesting process. Therefore extra care is needed when bottling your honey for showing.

To get the jars spotless, wash them with washing soda, as detergent can leave a film on the glass which can taint the honey. Let them drip-dry upside down and when dry buff with kitchen towel, especially the jar's shoulders. To reduce air bubbles, pre-heat the jars before

pouring in the honey. Slightly overfill each jar so that you can skim off any bubbles that rise to the top and still retain the correct weight. Carefully wipe and clean the rim to stop any honey trickling down the sides. Any bubbles that do appear, commonly round the shoulder of the jar, can be scraped away with a thin, bent piece of wire. Any that settle at the top can be skimmed off with a knife. Save a few extra jars at the time of bottling so that you can choose the best when the date of the show comes around.

There will also be strict rules on the size and content of the labels, so check the entry requirements.

There are different classes for showing honey according to its colour (light, medium or dark); what plant it derives from; and the form it takes, which includes:

- liquid honey (extracted honey with little processing)

Above Small, attractive jars make the honey an ideal gift for friends and family, or to sell at a farmers' market.

- set honey or creamed honey (blended liquid honey where the crystallization process has been controlled by careful stirring)
- comb honey (honey-filled beeswax comb cut off the frame)
- chunk honey (a piece of beeswax comb honey in a jar with liquid honey poured around it).

Honey judges are meticulous and have an array of tools to aid them in their search for the perfect entry. A large torch to shine through the jars, along with a magnifying glass, picks out any impurities that may be lingering in your honey; tasting rods check the flavour and texture; coloured glass identifies the grade of the honey; and a refractometer measures the water content. They also have a

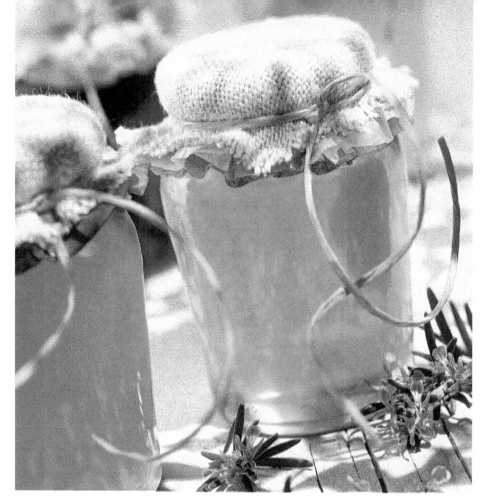

production, such as pasteurization
- the information needed on your labels
- the weight of honey you can sell
- the type of container and lids to use.

It is also likely that your region will have specific rules for handling and dealing with foodstuffs for public consumption, so do familiarize yourselves with the basic risks and practices involved by contacting your food standards agency. Your local council or national honey board will be able to advise you on specific criteria.

IS MY HONEY ORGANIC?

Honey can only be called organic if the hive is situated on land that has been certified as organic and since bees can forage quite a distance away from their hive, anywhere within at least a 5km (3 mile) radius of the hive must also be organic or uncultivated. If you don't adhere to those criteria, you can't label your honey as organic, so it is safer to say locally produced. That way people can choose to buy your honey instead of another variety produced further away that has added 'food miles'.

Honey is highly nutritious. It contains vitamins and antioxidants and is fat free, cholesterol free and sodium free. Local honey is also thought to be an antidote to hay fever. And urban, polyflower honey, made from the rich biodiversity of flowers and plants found in our towns and cities, is thought to be more flavoursome than honey from the more monoculture countryside.

weighing machine to ensure the weight is exactly as it should be. With their tools and experience, the judges will quickly whittle down the entries to a shortlist of around six jars that will be further examined for smell, taste, weight, clarity, texture and overall quality.

Judges and fellow competitors are a friendly bunch and will be happy to share their tips and techniques, so if your first year isn't a success, take their advice on board. Most importantly, don't give up.

Selling your honey

You may decide that you want to sell your honey at a farmers' market, at your local village fete or even to a local retailer. Whichever outlet you choose you will have to adhere to guidelines for selling food products direct to the public. Each country has its own rules which cover hygiene, food standards, health and safety and labelling. Honey is no exception.

Wherever you live, there are likely to be rules about:
- the water content of your honey, probably limiting it to under 19 per cent, since honey is liable to ferment when the water content is greater than that
- the processes you can employ in

Honey in folklore

In Greek and Roman mythology, ambrosia is the food of the gods and nectar their drink. Both are believed to be honey, which was thought to have an 'ethereal nature and origin'. As Aristotle said, 'Honey falls from the air. Principally at the rising of the stars and when the rainbow rests upon the earth.' The Roman encyclopedist Pliny the Elder saw bees as merely carriers of the food he called the 'sweat of the heavens' and 'the saliva of the stars'. Virgil said that bees 'partake of an Essence Divine and drink Heaven's well-springs'.

Throughout the ancient world honey was used to mark birth, marriages and death, and was a widespread medicine. Babies were fed it to ward off evil spirits and in Greek mythology baby Zeus was thought to have been saved from being killed by his father by eating honey fed to him by sacred bees or bee maidens. The Egyptian pharaohs used it to make offerings to their gods and were buried with it. Their funeral vaults, dating from 1450BC, feature wall paintings of collecting honey, jarring honey and baking honey cakes.

In the Old Testament there are numerous references to the promised land as 'flowing with milk and honey', which symbolized riches and plenty. But honey has also come to be a symbol of the sweetness of love. Cupid's arrows were sometimes said to be dipped in honey, while the Hindu god of love, Kama, carried a bow with a string made of bees. The word 'honeymoon' may come from the Viking custom of a bride and groom eating honeyed cakes and drinking mead for the first month of

their marriage, although it may just mean that the first weeks of marriage are the sweetest.

In the ancient world honey was used for embalming and its antibacterial and antiseptic qualities healed wounds, treated stomach disorders and cured coughs. There were even audacious claims that it could cure baldness and blindness. Today, taking local honey to treat hay fever is gaining credibility.

Ever since Cleopatra bathed in asses' milk and honey, its rejuvenating and beautifying powers have been legendary.

The oldest records of humans taking honey are in prehistoric cave paintings in Spain. One, at Cuevas de la Arana in Valencia, shows a man climbing up a cliff to rob a swarm of wild bees. It is dated around 15,000 years ago, just after the last ice age. Honey hunting continued into the ancient, medieval and modern world. The pioneering spirit of the honey hunters in the wild west is celebrated in US honey lore, since it is said that the mouths of the bees' caves were guarded by rattlesnakes, bats, scorpions, ghosts and Cherokee Indians.

Bees in winter

The winter months are the quietest for the beekeeper. The bees are huddled together in the hive keeping warm and feeding off their stores of honey. But before the cold weather sets in, they will need your help if they are to survive until spring arrives. As well as ensuring that their home is watertight, well ventilated and guarded against unwelcome intruders such as mice, you need to prepare a large meal to compensate them for the honey you have removed, and practise diligent pest control to prevent your colony being weakened by disease.

Preparing for winter

Late summer is the end of the bee's year. Even if the weather is hot and there are many later flowering plants in bloom the honey flow will have decreased, the colony's population will be falling and the workers will be evicting the drones. The queen will reduce her laying and the brood nest will be shrinking to a few frames.

Assuming you have removed the honey supers for harvesting, the hive will have shrunk to brood box height. Around 10,000 bees will live in the single brood box over winter. It will give the colony the warmth and comfort it needs to survive the cold months. But first they need a little help from you.

Below: A mouse guard fastened across the entrance to the hive during winter will prevent any small rodents from intruding into the bees' warm home.

Autumn feed

Remember that bees need to evaporate water off nectar to make honey? Well, the same applies to the sugar syrup you feed them. For this reason, you need to give them their autumn meal while the weather is still warm enough for this evaporation to take place. Even if it is hot and dry into early autumn and you think there is a nectar flow, it is still advisable to give them their sugar-syrup solution in the first few weeks after your final summer harvest. If you leave it too late and the weather has turned cold they will take it, but will not be able to process it and store it for winter.

How much do you give them? It is thought that an average colony of bees needs around 18kg (40lb) of stores to see them through the winter. So you first need to assess how much honey they have left in the hive. Each standard national brood frame holds just over 2kg (5lb) of honey. It is likely that your bees will have four outer frames full of honey and that the rest will be a mixture of brood, pollen and honey stores. So they will probably have just over 9kg (20lb) of honey, about half their requirement. This means you need to feed them around 9kg (20lb) of sugar to supplement their stores.

Make up a strong sugary solution with a 2–1 ratio of sugar to water. It can take a while to dissolve such quantities of sugar even in warm water, so prepare your solution in the morning and feed it to the bees in the late afternoon when the foragers have returned to the hive. The best way to do this is with a rapid feeder (see 'The kit') and you could also take the opportunity to supplement the feed with some medication to treat for Nosema (see 'Prevention and cure').

A colony of 10,000 bees can empty a 4.5 litre (1 gallon) fast feeder in less than 48 hours if they need the nourishment, so be ready to give them more. Don't worry about overfeeding – they will consume only what they need. There are no reports of obese bees. They will take the food down into the hive and use the available cells for storage. When they have stopped consuming the liquid, you can remove the feeder.

Leaving supers

It is also possible that rather than taking away most of the bees' excess stores of honey for your own use and replacing it with sugar

Left Your winter hive will contain just the brood box, in which around 10,000 bees will huddle throughout the cold months. Preparing for winter is crucial for the beekeeper, and includes an autumn feed, pest control and ensuring that the hive is dry.

solution you may decide to leave the bees their own honey that they have been collecting during the season. A full super is between 20–30lbs of honey which, when added to the honey that is in the brood box, should be enough stores to see the colony over the winter. The choice to leave the bees their own honey or to take it off and replace it with sugar syrup is up to you. It will depend on a variety of things like the amount of honey you harvested, but it is also down to fundamental questions such as why are you keeping bees?

Up until recently most hobby beekeepers kept bees for the honey and the old beekeeping books are written with one thing in mind; how to maximise the amount of honey your busy workforce of bees can make. But now with the rise of the eco-friendly beekeeper who is keeping bees in their urban backyard to aid pollination, honey isn't always the priority. If this is the case, then by all means leave your bees more of their own honey to eat over the winter. But don't feed honey to your bees that is from another hive. Only give bees their own honey; the honey can harbour bacteria that cause foulbrood disease (see 'Prevention and cure') and we don't want our colonies getting ill.

Winter feed

We can also feed our bees a winter supplement of sugar in the form of bakers fondant (sugar candy). This solid feed can be purchased from catering supplies, local bakers, beekeeping suppliers or you can make your own. Wrap a block of candy in greaseproof paper or newspaper, make a hole on the underside and place it on top of the crown board's feeding hole, leaving some room for ventilation. The bees will come up through the hole to eat it. Feeding candy is a good precautionary measure against your bees starving and it may help to ensure the colony comes out of the winter and through the early spring.

SUGAR CANDY

2lbs sugar
½ pint or less water
1 tablespoon liquid glucose
• Heat water, sugar and glucose together gently until sugar dissolves then bring to 115°C.
• Place pan into a sink with cold water (to cool mixture quickly) and start stirring with a wooden spoon.
• Keep the mixture moving: when it gets too hard to stir and has cooled enough to work with hands, tip out onto a worktop/silicon sheet splashed with water and knead until cool.
• Fill small tubs and use when needed.

Hefting the hive

Knowing how much stores your bees have is important for your piece of mind and also for the colony's welfare. One practice that is worth getting into the habit of using is hefting the hive. This is simply lifting the back of the hive until the back legs of the stand just come off the ground. If you practise this over time you will be able to tell if your hive is heavy or light with stores. You can be more precise when hefting if you use a luggage scale. Remember a hive wants 40lbs of honey for the winter. Eventually hefting will become part of your inspection routine and it will give you the first clue as to the state of your colony. If it is light you may need to give them extra sugar.

Varroa control

Another important job before the cold autumn weather arrives is to administer some form of varroa mite control. Unfortunately the varroa mite is widespread throughout most of the world and is likely to occur wherever there are western honey bees. To contain these debilitating parasites there is a range of different methods and treatments. The reason for treating the colony in the late summer is that we want our winter brood and bees to be as free of varroa as possible giving the colony the best chance of survival. We discuss Varroa in more detail in 'Prevention and cure'.

Below *Even when it's freezing outside, bees keep warm in the hive by clustering together.*

Keeping the bees dry

Bees can handle the cold well by keeping warm as a cluster, but damp is something they don't like. Who does? Inside a hive it can get quite humid as the bees expel the water content of the honey and this water can collect on the solid floor. The best way to beat the damp is to give the hive adequate ventilation. When using solid floors in the hive this can be achieved by placing a match under the corner of the crown board, allowing stale, damp air to escape through the small gap and the ventilation screens in the roof. Now, as part of your integrated pest-management system to control the varroa mite, your hive should have a screened mesh floor. If you

SCORCHING

The woodwork of beehives is hot and humid – perfect conditions for many unwanted organisms. The most effective way to clean and sterilize anything you are going to reuse and to keep it pest- and disease-free, is to scorch the woodwork with a blowtorch – the type commonly used by plumbers. Do this after the spring cleaning in your second and subsequent years. Thoroughly scorch the insides and outsides of the brood and super boxes, floors, crown boards and queen excluders, paying particular attention to the corners and any small cracks. Don't burn the wood, just give it enough heat to toast it. This operation is especially important if you have more than one hive and you interchange hive bodies and parts.

leave this on during the winter (without the monitoring tray) it will give the bees adequate ventilation and it won't let water accumulate.

Guarding against intruders

During the winter mice are always on the look-out for a cosy spot to bed down, and a hive can make a perfect home for them, so you will need to reduce the size of its entrance. Remember that a mouse can get through very small holes. There are a number of different entrance blockers – ask your local suppliers to recommend their favourite. One of the most effective is a simple galvanised strip with two rows of 10mm (½in) holes, called a mouse guard, which is fastened over the hive entrance with pins. Occasionally during the winter check that the holes are not blocked so the bees don't get trapped inside the hive.

Depending on where your hives are, woodpeckers, badgers, skunks and raccoons may also try to get inside that cosy, honey-laden home either for food or for comfort, so precautions are needed to stop them if they are common in your area. Wire-mesh screens around the hive should do the job, but if you live somewhere where there are bears, more substantial defences such as an electrified fence may be necessary.

What is happening in the hive

In mid-winter, the colony will have reduced in number to around 10,000 worker bees – a cluster large enough to maintain the heat in the hive but small enough to live off of the amount of honey stores. The queen will take a rest from laying, though she will begin again shortly after the winter solstice.

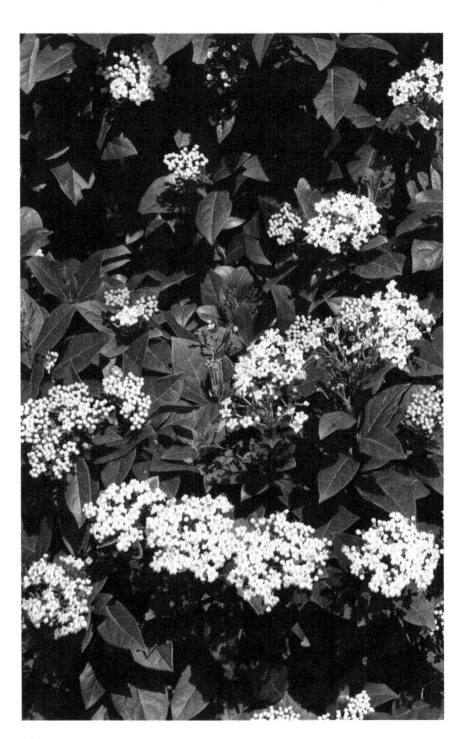

Above *On a mild winter's day there is little food available for foraging* **bees** *except a shrub such as viburnum tinus with its tiny white flowers.*

The bees cluster together in a huddle, taking turns to be on the colder edges, and feed on their stores. Since the workers are not out foraging, most of those that were born in late autumn will live through to the spring. There will be some deaths, however, and the bodies will be dragged out of the hive, so it is not surprising to see some corpses on the floor outside the hive entrance throughout the winter.

On a cold, sunny day some bees may take a short trip out of the hive to stretch their wings and take the opportunity to relieve themselves, since they keep the inside of their home very clean. They won't go far and may even collect some fresh pollen if there is any nearby. It is not unusual to see bees coming into the hive with pollen on their legs on a warm winter's day. This is a good sign, since fresh pollen is an indication that the queen is laying and there is hungry larvae being fed.

If all goes to plan, the mass of bees will work its way around the hive, eating the stores of honey and begin to increase in population as the winter ends and spring arrives.

Looking in the hive

Sometimes however the cluster of bees doesn't manage to work their way around the comb to get to all the available stores. If the bees start from the middle and work out to the front of the hive eating all the stores they in effect cut themselves off from the honey at the other end of the box and if it is too cold they might get stranded at the far end and not survive even though there is honey in the nest.

If you suspect this may be the case, you should open the hive very quickly – remember bees don't like the cold air – to move frames of food around. If there are frames of stores at the far end of the box move them nearer to the cluster of bees. This should only take a minute or two. As long as you are quick and they have the food the colony will soon warm up again. We would rather make sure the colony has access to a food store than to leave them to starve.

Stock check and reading

Take advantage of the quiet winter months to stack and store away any frames and supers that are not being used. Store these outside in the cold, but keep them dry and ventilated. The cold will keep away the wax moth.

This is the time to make a note of the equipment you have, to give it a thorough clean, see if any needs replacing and decide whether you will need extra equipment for the second year of beekeeping. Would your next harvest be easier if you invested in an extractor? What about a more light-weight bee suit? You might even be thinking about whether to increase the size of your apiary – two hives are not much more work than one. It is also worth investing in a second hive in preparation for swarming and increasing your apiary.

Winter is also a good time for the bee-keeper to go on a theory course at their local association and read the magazines and books. There is always more to learn about bees.

Towards the end of winter make sure your equipment is ready for the coming season. Be prepared since spring comes quickly and the colony will be making plans to swarm well before the weather is warm enough for us to be comfortable in short sleeves. Remember the bees work to the sun and by the spring equinox (21 March in the northern hemisphere) the flora begins to flower giving the bees the food they need to grow, expand and procreate. The colony expands fast after its winter slumber and may decide to swarm much earlier than you might have expected so be ready to start your inspections for queen cells just after this time of year when it can still feel a bit chilly.

Occasionally during the winter check that the holes are not blocked so the bees don't get trapped inside the hive.

The second year

When the bees begin foraging regularly for food again, their appearance heralds the beginning of spring as surely as the crocuses and the daffodils. Start your new season of beekeeping by giving the hive a thorough spring clean as soon as the days are warm enough for you to open it up. Your bees have been cooped up in their home for months, so they will appreciate new comb and a new brood box. Bees may swarm in the queen's second year. Regular inspections of the hive will allow you to judge whether or not they are planning to do this and to implement swarm prevention or control. Swarming is the bees' natural way of propagating the species and is nothing to be scared of. It may alarm the neighbours, but a swarm is often easy to collect and is a free and simple way to increase your apiary.

The start of the second year

Hopefully, with your care and support, your colony will have made it through the winter and you will be able to enjoy your second year as a beekeeper.

Early spring

Don't think that you can sit back and relax just yet. The arrival of the first spring flowers can signal a particularly perilous time for the colony. The warm weather might bring some welcome flows of nectar and pollen and this will set in motion the bees' natural tendency to increase the size of the colony, making them think the higher temperatures are here to stay. Often, however, especially in temperate regions, a period of cold, rainy weather can quickly follow. This will prevent the bees from foraging and they will have to rely on their dwindling stores to feed the increasing population of brood and young bees. This is the time when the bees are most likely to run out of stored brood and starve. For this reason, you must be prepared for any action you need to take to help your colony during those few weeks. If the weather does change for the worse, feed the bees a strong sugar syrup solution in a contact feeder and refill it if needed. It is better to be safe than sorry. As with the winter feed, you can't overfeed them: if the bees need the food they will take it down into the hive.

Too late

The mistake many novice beekeepers make in the second year is to wait until the temperature is warm enough for us to be comfortable in short sleeves before they start to think about their beekeeping. This is too late. Remember bees work to the sun so by the spring equinox (21 March in the northern hemisphere), your bees could be getting ready to swarm. One of the skills a beekeeper needs is to be able to anticipate events. There are many instances when you will need to be one step ahead of your colony to ensure you are able to help them or to manage their swarming tendencies. So in the very early spring when the snowdrops and crocuses are out, even though it may still be cold and the sun may be behind the clouds, this is when you need to start your beekeeping for the second season.

Spring clean

One of the main jobs in early spring is to get your bees onto fresh brood comb. The purpose of this is to replace the old brood comb with new fresh comb, reducing the risk of disease and swarming while at the same time you give the bees' home a freshening up.

This spring clean mirrors what a colony in the wild would achieve with the help of the wax moth. A natural colony would migrate up the comb to the stored honey at the top of the nest. This leaves the old brood comb exposed and readily devoured by the wax moth. In the spring when the colony expands again, the old comb has gone and the bees make new comb for the queen to lay eggs in. When you keep your bees in the hive the wax moth doesn't perform this function, therefore the beekeeper needs to replicate it by taking away the old comb and replacing it with foundation for the bees to draw out.

You will need to add a new brood box filled with new frames and foundation on top of the old brood box. The bees will draw out the foundation and this new drawn-out comb is preferred by the queen and she will happily lay her eggs there. Once she has vacated the old box and moved upstairs, and the brood below has developed into adult bees, you can remove the old frames and brood box. The process also combines the opportunity to replace the stand, cover board, entrance block and floor with cleaned (sterilised) or new ones.

The brood comb change takes about a month to complete and you will want to have the procedure finished before the bees would normally begin their swarm preparation. So start early in the spring, even if the air is a bit chilly, aiming to complete the comb change before the swarming season begins. To carry out the process you need only visit your hive three times for half an hour for each visit.

For the first visit, wait for a dry day. It may be cold so it will be good to act quickly. You are not going to disturb the brood nest and as long as the colony has enough food it will soon replace the lost heat. You will need to have a new brood box filled with frames and foundation, a new crown board and a feeder with sugar syrup solution. Light your smoker, put on your protective gear and give the hive a small puff of smoke a few minutes before you open it up. When you open the hive you should see the bees occupying four or five frames, not necessarily in the middle of the brood box. These will be the frames that have brood on them, with nurse bees attending to the larvae. The other frames are likely to be empty but some will have stored honey. These broodless frames are the ones you want to remove.

Start at the furthest frame from the cluster and inspect that frame. If it is empty or has a few cells of capped honey, remove it and place it to one side. Carry on looking through the frames until you come across a frame which has pollen on one side – this will probably mean that the other side has brood. If it has, put it back into the brood box in the same position. Now start at the other end of the box and do the same. You will end up with around five frames of pollen, brood and bees which haven't been disturbed. Slide these five frames to the centre of the brood box and on either side return a frame which has some

honey stores on it. You will now be left with seven frames centred in the brood box.

Take your new brood box full of new foundation and put this on top of the old box with the frames lined up with each other. Now put on the crown board and feed the bees the sugar syrup solution. Drip some of the solution into the hive to let the bees know there is food at the top. This feed gives the bees the energy to draw out the new foundation which will encourage them into the new box above. Close up the hive.

Once the bees have started to draw out the foundation the queen, who seems to prefer fresh comb, will move up into the new brood box and start laying. The nurse bees will follow to feed the brood and the migration to the top box will take place.

After a week, plan your second visit. On opening up the hive and looking into the top brood box you will hopefully see a cluster of bees drawing out the foundation and looking after the newly laid brood. Then introduce the

Above *It is a good idea to give your hive a spring clean as soon as the days are warm enough for it to be open. Old foundation can get damp over the winter, and new, fresh comb can reduce the risk of disease.*

Left *To start the spring clean have five frames at the centre of the brood box with pollen, brood and bees, and on either side a frame that has a lot of honey.*

One of the triggers that may lead to swarming is a lack of space so be sure to have your supers ready to add to the hive.

queen excluder between the two brood boxes and make sure the queen is in the top box. If you are lucky she will be up there but you may need to find her by looking through the bottom box. Pick her up gently and drop her into the top box. She will soon be followed by her attendants and the workers who will continue feeding the brood and drawing out the rest of the foundation.

If on your second visit the bees have not yet drawn out foundation in the new brood box, and so there is no brood, you will have to wait another week before you can add the

queen excluder.

Your final visit will be 21 days after the queen has been isolated in the top box. All the brood in the old bottom brood box will have hatched and we can now finish the spring clean. This will be your final visit where you will need to remove the old brood box and frames below, replace the old floor, entrance block and crown board with new (or cleaned) ones. Check that the colony still has room to expand. If not you will need to add a super.

You can reuse the old brood box after cleaning and scorching but dispose of the

Left *Honey bee on an alder catkin collecting the early spring pollen.*

Left *Carrying out an inspection.*

frames after melting down the wax, in case they are harbouring any disease. As a result of your spring clean, the bees can enjoy a hygienic home with the reduced risk of infection, and they also have lots of clean foundation which will keep them busy and thereby reduce their swarming tendencies.

Keep a close eye on the new brood box to ensure that the bees don't run out of space. Once a good spring flow is available the colony will expand very rapidly and will need extra room to deposit the nectar. One of the triggers that may lead to swarming is a lack of space so be sure to have your supers ready to add to the hive.

Adding supers
By now the days will be getting sunnier and warmer and the bees will be out in force foraging nectar from the blossoming fruit trees, and flowering shrubs and plants. Since nectar has a much higher water content than honey it takes up a lot more room in the hive so the bees will need plenty of space to deposit it. Add a super as soon as you see bees drawing out the foundation on the last couple of frames. Always have a spare super box complete with frames ready for your colony since they may need it in a hurry.

Weekly inspections
Your weekly inspection will start again in earnest in the spring. You will be looking for signs of swarming, space requirements and the health of the colony.

The reason for weekly inspections is that they give you the opportunity to notice queen cells developing. It takes 8 days for a queen cell to be capped which is the moment the swarm leaves the nest. By carrying out inspections every 7 days you will have time to spot swarming activity and then do something about it. If you are going to prevent unwanted swarms you may need to carry out these weekly inspections even if it is cold or raining, though it is advisable to wait until it has stopped raining if you can. Your bees will appreciate it.

If all goes well the colony will continue to expand. They will forage on the spring crop and collect their valuable food. Second year beekeeping is exciting and can be full of surprises. You will enjoy seeing your colony flourish and the foragers bringing back pollen in a rainbow of colours, each one telling you they have been pollinating the vast array of different plants, trees and flowers within their flying radius.

Swarming and increasing an apiary

The honey bees' natural way of propagating is to swarm, creating two colonies from one, which effectively multiplies the species. The old queen and around half the colony leave and settle in a new location. The bees that are left behind raise a new queen, look after the larvae and unhatched pupae and grow to the size of a full colony.

Swarming is healthy for the bees, but not always welcomed by the beekeeper. It not only reduces the amount of honey that a hive produces, since the deserters take a lot with them and there are fewer workers to forage for food, but it is likely to scare the neighbours. A black cloud of 25,000 bees is an alarming sight for most people.

Swarming normally occurs in late spring or early summer. The triggers are:

- overcrowding, which means the queen has too little laying space
- old, unusable comb
- an ageing or diseased queen.

How to collect a swarm

Although swarming can look menacing, as long as everyone stays calm it is a fascinating event. If you are lucky and the swarm lands somewhere accessible, then you can collect it and increase your apiary for free.

> Swarming is healthy for the bees, but not always welcomed by the beekeeper. A black cloud of 25,000 bees is an alarming sight for most people.

When a swarm leaves the hive, usually before midday, the bees fly around in a haphazard fashion for some 20 minutes before massing into a black cloud and finding a temporary place to rest, such as a tree branch, within 50m (55 yards) or so of the hive. Hanging from the branch, they surround their queen in the shape of a rugby ball and wait there for anything from an hour to a day until their scouts have located a viable new home. This sojourn is your opportunity to collect and rehouse a swarm. Filled with three days' worth of honey they are a pretty soporific bunch, but you should still wear your protective gear and have your smoker at hand.

The principle of collecting a swarm is pretty straightforward. You want to get the bees off their temporary landing spot and house them in the new brood box. You do this by either giving the branch they are on a short, sharp shake, or cutting it, so it and the bees fall into the container that your assistant (also wearing protective gear) is holding just below. If the swarm is on a gatepost or fence, you can use a bee brush or your gloved hand to dislodge them; alternatively, if you can get the receptacle above them, you may be able to encourage them into it by giving the cluster a gentle puff of smoke from below.

Once you have them, cover the container with your sheet and take them to their new home. The bees' natural instinct is to follow the scouts and the queen into the new home and you can encourage this by placing them outside the hive and letting them find the entrance. Start by positioning a piece of board (1.2m x 60cm/4ft x 2ft) on the landing board and let it slope down to the ground. Cover it

with your white sheet. Make sure the sheet and board are touching the entrance but not blocking it. If you now turn over the container of bees and dump them on to the sheet, they will crawl up it (they like walking uphill) to the dark entrance. Once the queen is inside the rest of the swarm will follow. It may take 30 minutes for the bees to walk up the sheet and into the hive. This process lets them think that they found their new home naturally and enables them to settle in that bit quicker.

Alternatively, you can just turn the container upside down over the new hive and shake the bees into the new brood box, complete with frames. Put the roof on the hive and leave the skep or box near the entrance. The bees inside the hive will be wafting their

EQUIPMENT

The equipment you are likely to need:

- **a pair of secateurs or branch cutters to cut the branch the bees are dangling from**
- **a receptacle for the swarm, such as a skep, cardboard box or nuc box**
- **a step ladder if the swarm is high up**
- **a sheet to cover the receptacle once the bees are inside**
- **an assistant, as it's difficult to reach up and cut a branchful of bees while holding a box underneath to catch them**
- **If you are planning to keep the swarm to increase your apiary, have a brood box, floor and frames in a suitable location so you can transfer the swarm once you've caught it.**

Above *If the bees swarm on a gatepost or fence, you can use a bee brush or your gloved hand to dislodge them. Alternatively, if you can get the receptacle above them, you may be able to encourage them into it by giving the cluster a gentle puff of smoke from below.*

Right *Swarms of bees can look menacing – especially to people who are not used to them – but they are simply the bees' natural way of propagating. It is also a way of increasing your apiary, for free!*

scent into the air and the stragglers will soon find their way to the new home.

You now have a second colony of bees. If you decide to keep them, rather than give them to your local organisation or another beekeeper, be aware that they expand very quickly, so be ready to add a super.

If you don't feel comfortable, you don't have assistance or the swarm is inaccessible, contact your local beekeeping association which will have experienced collectors on standby to come and take the swarm away. It will be given to one of the people who have put their name down on a swarm list in the hope of increasing or starting an apiary.

Cast swarms

It is likely that if a colony swarms once the remaining bees will swarm again, and possibly a third or fourth time.

When a colony swarms it leaves behind half the bees, brood at all stages and a new queen in the making. It ensures that there is an emerging queen by rearing a number of them, maybe a dozen. These unborn virgins will emerge nine days after being capped over, which is usually the day the first or primary swarm leaves the colony.

The first virgin queen to emerge after the primary swarm has gone may decide to throw what is called a 'cast' swarm. She takes half the remaining bees, probably around 13,000, and looks for a new home. Those who are left will wait for the next virgin queen to emerge from a capped queen cell and she may cast again by taking half the remaining bees away. This could continue over a 10–12 day period as the queens are born, until the hive has been vacated completely.

Left *A swarm will often collect on a tree branch. They hang from the branch, surrounding the queen and wait there until the scout bees have located a suitable new home.*

Swarm prevention

There are many different ways to stop your bees swarming and none of them are foolproof. But you can use your knowledge of bee behaviour to give them the impression they have swarmed, which satisfies their instinct but keeps them in your apiary.

Because bees often swarm when they are overcrowded, an ongoing swarm-prevention technique is to make sure that the queen has enough brood area to lay her eggs. This means plenty of empty comb, which is why it is a good idea to replace old brood comb with new comb each spring. The colony also needs plenty of room to store nectar, so add supers to the hive before they are needed rather than wait until the storage room has run out.

Further techniques include replacing queens every year or second year. A younger queen's pheromones are strong, so they suppress the queen-rearing instincts of the colony, making it less likely to swarm.

Queen cells

Can you tell when a colony is going to swam? Since it will only do this when it has reared a number of virgin queens to take over the remaining colony, you need to know what queen cells look like. If they are occupied by a larva this indicates that a swarm is about to happen.

A queen cell is a cone of wax about the size of an acorn, usually hanging downwards from the bottom of the frames, though it can also hang downwards in the face of the frame. It is common to find an empty

tip
A colony will only swarm when it has reared a number of virgin queens to take over the hive.

queen cup which is really a queen cell that is waiting for an egg. If you see a lot of queen cells, your colony has probably started preparations for swarming.

A queen cell comes to a narrow point. If this point has a hinged flap that is open, and by looking into the cell you can see that it is empty, then your colony has raised a virgin and she has emerged. This will probably mean that your bees have already swarmed and you missed it. If this is the case and there are eggs and larvae in your colony, you have a new, young queen in her first year. Let the colony get on with things but make sure they have enough room throughout the year to prevent further swarming. If there is one open queen cell but other closed queen cells you can assume the virgin has just emerged and is in the hive. You will have to destroy the other queen cells to stop the hive throwing a cast swarm. A hive can only support one queen, so if another virgin queen is born, the colony could swarm again.

However, the queen cells are not empty but have larvae in them, the colony is getting ready to swarm and will be off as soon as the queen cell is capped. This happens 5 ½ days after the egg has hatched into a larva, so you have time to prevent the swarm leaving. You do this by splitting the colony, which will give the bees the impression that they have swarmed and curb their instincts for swarming again. One of the many ways to do this is to create a nucleus.

Creating a nucleus

You will need a nuc box or a spare brood box with five brood frames, a floor and cover board.

1. Look through the hive and find a frame with pollen and honey. Place this and the bees into the nucleus box.

> You can use your knowledge of bee behaviour to give them the impression that they have swarmed, which satisfies their instinct but keeps them in your apiary.

2. Find the queen and the frame she is on and place this into the nuc box too. This frame should have a little brood at all stages. If you see any queen cells, remove them.
3. Find another frame with stores of pollen and honey and put this into the nuc box, again removing any queen cells.
4. You need some bees in your nuc box, so shake and brush all the bees off two more frames into the box.
5. Fill the remaining space with empty frames.
6. Close the box and position it on a new stand at least 1m (3ft) away from the original hive.
7. Look through the original hive and destroy any queen cells that are about to be capped, leaving some queen cells that have young larvae in them.
8. Close up the remaining frames in the original hive and fill the spaces with new frames.

   You now have the original hive with queen cells and no queen, and a nuc box with the old queen, some brood, nurse bees and some stores.
9. In one week return to the original hive and destroy all but one of the capped queen cells. The best way to do this is to place the first frame you find with a queen cell to the side, before destroying it – just in case the first one you find is the only one – and then destroy all subsequent cells.

10. Once you are sure there is only one queen cell left, put it back in the hive, close the hive and leave for four weeks. Don't be tempted to open the hive too soon as virgins and newly mated queens are easily disturbed. When you return for the inspection the virgin should hopefully have mated and begun to lay.

The nucleus can be left to build up into a new colony or, if you don't want to expand your apiary, the bees, without the queen, can be united with the original colony when its new queen is settled in.

Has your hive swarmed?

Bees swarm during the day, when you are quite likely to be out. When you come to inspect your colony you may notice that they are a bit more agitated than usual and buzzing loudly. This may be an indication that they do not have a queen who is laying eggs. Look through the colony to see if there are any empty queen cells. If there are, then your bees have probably swarmed. To make sure, check for eggs and young larvae. If you can't find these, then your laying queen has gone. There will probably be a virgin queen running around the frames. She will be difficult to spot since she is smaller than a mated queen (her spermatheca is not full and expanded). Try to find any other queen cells and destroy them. A cast swarm will only happen if there are queen cells in the colony.

Close the hive and leave it for four weeks to give the virgin time to mate and start laying. When you inspect the hive after a month you should find new brood, which indicates that the colony is back on track.

Bee products

Not only will your bees give you honey, they will also produce beeswax with which you can make the most heavenly scented furniture polish and candles, and natural cosmetics that are thought to have special health-giving powers. After extracting the beeswax, you can have hours of fun with all the family learning to create different types of candles in all shapes and sizes, or making a variety of creams and ointments. It can be a bit fiddly at first, so don't be put off if you don't get it right straight away. Your patience will soon be rewarded and you will enjoy yet another contribution from your bees. Once you have mastered these crafts your artistic work will make ideal gifts for all occasions.

Beeswax

Your wax will come from the cappings that you have saved during the honey harvesting as well as scrapings of comb that have come off the frames and hive during the course of the year. In your second year you will have more wax as you replace some of the older frames with new foundation. This wax will have to be melted and filtered to give you clean beeswax for your candles.

Extracting beeswax

As your apiary grows you may wish to invest in a solar extractor, which is a clean and labour-free way of getting wax off the frames. It consists of a shallow box, into which you place the frames, with a glass lid which traps the sun's heat, causing the wax on the frames to melt. The wax is collected in a tray after passing through a filter which collects the debris.

If you don't have a solar extractor you'll have to melt the beeswax by hand. Utensils that come into contact with wax will be covered with the stuff, so have dedicated ones just for wax handling.

You will need a double-boiler or bain marie. This is a pot within a pot that allows you to heat the wax indirectly by boiling water in the outer pot while using the inner one for the wax. You can improvise by placing a small pot into a larger pot of boiling water, or a glass or stainless steel bowl over a large container of water.

You will also need a metal strainer and collecting pan for coarse straining, and nylon tights or muslin to fine-strain the wax.

Warning: Never melt wax directly on a flame or in a microwave. It gets hot too quickly, and can easily burn and ignite. Always use the double-boiler method. Make sure all workbenches are covered with old newspaper to catch the inevitable drips.

1. Heat the water in the larger, outer pot and place the collected wax into the inner pot.
2. Add some water to the wax so that it has something to float on and debris contained in the wax will sink to the bottom.
3. When all the wax has melted, pour it through the metal strainer into the

tip
Beeswax obtains its colour and fragrance from the propolis and pollen brought into and stored in the hive.

> ## BEESWAX FACT BOX
>
> **Beeswax is a by-product of the honey bees' honey consumption. It takes about 4kg (8 1/2lb) of honey to produce 450g (1lb) of beeswax.**
>
> **Worker bees aged between 12 and 18 days secrete beeswax in the form of a scale about the size of a pinhead from wax-secreting glands under their abdomen. It's estimated that it takes about 500,000 scales to make 450g (1lb) of beeswax.**
>
> **The secreted wax is tasteless, odourless and almost colourless. It obtains its yellow colour and its fragrance from the propolis and pollen brought into and stored in the hive. Beeswax has a melting point between 63 and 65°C (146 and 149°F).**

collecting pan. This gets rid of larger bits of waste such as old silk cocoons, bee parts, propolis and even the odd bit of foundation wire.

4. Give the warm inner pot a quick clean with some kitchen towel and return the strained wax to it. Melt the wax for a second time and strain it through the tights or muslin to get rid of finer particles such as pollen.
5. Pour it back into the inner pot with enough rainwater to keep the wax off the bottom. Heat it for a third time until the wax has melted again. Now let it cool for a day or two. It shrinks slightly on cooling and breaks away from the edges of the pot, leaving a cake of yellow wax floating on top of the water.
6. There may be some bits that need scraping off the underside of the wax to leave a perfect block of clean wax. Put aside some beeswax in a small plastic container for use around the home. It makes a marvellous polish that not only seals the wood, but creates a high sheen and gives off a lovely aroma.

Beeswax wood finish (for bare wood)

You will need: 55g (2oz) beeswax, 300ml (½pt) turpentine, bain marie

1. Melt the wax in the bain marie.
2. Remove from the heat source and pour in the turpentine.
3. Return to the heat and give a gentle stir.
4. Cool, and spoon into sealable containers.
5. Warm slightly before applying to bare wood.

Apply with a soft cloth, allow to dry, then buff with a wool cloth. This product can also be used on leather.

Beeswax wood conditioner (for finished wood)

You will need: 55g (2oz) beeswax, 600ml (1pt) turpentine, 2 cups water, 25g (1oz) soap flakes

1. Shred the wax into the turpentine, cover and leave in a warm spot such as a sunny windowsill for a few hours, shaking from time to time until the wax has dissolved.
2. Bring the water to the boil, add the soap and stir until dissolved.

3. Let this cool and add it to the wax/turpentine mixture, stirring briskly until an emulsion forms.
4. Pour into sealable glass bottles and shake well before use.

Dubbin (traditional waterproofing for leather boots and shoes)

1. Stir 110g (4oz) beeswax and 55g (2oz) of palm oil together in the bain marie until the wax has melted.
2. Let it cool slightly before pouring into a jar.
3. Warm slightly before applying to leather.

Above *Beeswax is a highly-prized polish for bare and finished wood, which gives a high sheen and a lovely smell.*

Other uses for beeswax

Filling in cracks and holes in furniture; rubbing on thread for easier needle threading and sewing; lubricating the runners on sticky wooden drawers, saw blades, woodworking tools and zips on clothes. Fishing lines float better with a coating of wax.

Beeswax candles

There is something special about a soft glowing light from a candle enhanced by the delicate perfume of beeswax. Making your own candles is a skill that requires time and patience to master, but if your first efforts aren't perfect, you can always melt them down and try again.

Moulds

Moulds can be made from glass, metal, plastic and silicone and are available from craft shops and online stores. In your first year, when you may have little wax, try making tea lights using old tea light containers. You can also use old yoghurt pots.

The wick

The secret of a good candle is in the wick. You need the correct size in relation to the diameter of the candle – too small and the candle won't burn, too large and it will smoke. Beeswax candles need a wick about twice the thickness of that for paraffin candles. If you are buying wick from a beekeeping supplier they will sell it for beeswax, but a general candle-making shop will probably only sell wick for paraffin candles. A bit of trial and error will get you the right wick-to-candle ratio.

The wick then needs to be primed so that it is easier to light and thread through the hole in the candle. Simply cut it to length and immerse it in some hot wax. Once the wick has had time to absorb the wax, remove it, stretch it straight and place it flat to let it dry. This will only take a few minutes.

You will need: a yoghurt pot mould, wick, molten beeswax, scissors, a little dishwashing liquid or glycerine

1. Make a small hole in the bottom of the pot and thread the wick through the hole. To straighten it, tie a knot in the hole end and pull it tight. Wrap the other end around a toothpick or pencil and lay this across the opening, making sure the wick is centred.
2. Take your block of clean wax and melt it in the clean bain marie.
3. Line the inside with the dishwashing liquid or glycerine, which will allow the candle to break free from the mould once dry. Silicone rubber moulds do not need this.
4. Block the hole in the base with a little mould seal (Blu-Tack) to stop the molten wax escaping from the hole.

5. Position the mould on a tray, slowly pour the molten wax into it and let it cool for 24 hours.

6. Cut the knot that was holding the wick in place and you should find that the wax has shrunk and falls easily out of the mould.

7. Flatten the bottom of the candle by heating the double boiler and rubbing the candle on the bottom of the inner pot until it has a flat and smooth end. Your candle is now ready for lighting.

Foundation candles

The foundation in the frames can also be used for making home-made candles. These are simple to make, burn well and are a great alternative to moulded candles when you run out of supplies of wax cappings.

The foundation sheets must be the unwired variety. They are sold by beekeeping suppliers and come in a huge variety of shades and bright colours for candlemaking.

1. Foundation works best when it is warm – it gets brittle when it is cold – so if necessary place it in a warm spot to soften it.

2. Measure the wick against the long edge of the foundation and cut it about 15mm (½in) longer, leaving enough sticking out the top for lighting.

3. Lay a sheet of foundation on a flat surface and place the wick along the edge to be rolled. Fold the foundation tightly over the wick and press to create a straight edge.

Now continue to roll the wick, like a Swiss Roll, keeping the cylinder straight and tight.

4. When the whole sheet is rolled, stick the loose edge down by gently rubbing your finger along the edge. You should now have a pure beeswax candle ready for lighting.

Variation: An attractive tapered candle can be made by using sheets of foundation that have been cut diagonally. When folding, keep the longer, straight edge as the base and the candle will roll into a conical shape. To add to the fun, you can decorate your candles by adding shaped bits of coloured wax. Cut out stars from foundation sheets, for example, and press them on to the candle. Use a little molten wax to glue them on if necessary.

Below *Foundation candles are easy to make using different coloured sheets, and make ideal gifts. Because no molten wax is involved, they are also perfect for children to make.*

Cosmetics

Beeswax has been an essential ingredient in beauty products for centuries, sought after for its natural softening and protecting qualities for skin. Balms, creams and ointment can all be made quite simply in the kitchen by mixing little more than essential oils and water into your own beeswax. Craft shops stock a range of attractive containers that you can use for homemade cosmetics.

Beeswax lip balm

This is an amazingly easy recipe.

You will need: 30ml (2 tbsp) grated beeswax, 20ml (4 tsp) almond oil, 10ml (2 tsp) honey, a few drops of an essential oil (optional), double boiler, small containers.

1. Put the ingredients into your double boiler and simmer over water until they have melted.
2. Remove from the heat and stir together.
3. Add a few drops of your favourite essential oil if you like.
4. Pour the liquid into small containers.
5. When the mixture is cool put the lids on.

Milk and honey bath

This is a modern take on Cleopatra's bath of asses' milk and honey.

You will need: 90g (3oz) powdered milk, 60ml (4 tbsp) honey, bowl.

1. Mix the milk and honey into a paste.
2. Dissolve in a warm bath as you run it.

Body lotion

A light lotion with a fragrant smell that makes an ideal present.

You will need: 300ml (½ pint) aloe vera gel, 5ml (1 tsp) lanolin, 5ml (1 tsp) pure vitamin E oil, 100ml (3½ floz) coconut oil, 15g (½ oz) beeswax, 225ml (8 floz) almond oil, up to 7.5ml (1½ tsp) essential oil of your choice (or more to prolong the scent), blender, double boiler, glass jars.

1. Put the aloe vera gel, lanolin and vitamin E oil in a blender.
2. Put your coconut oil and beeswax in the double boiler and heat until fully melted.
3. Into this melted wax stir the almond oil, reheating if necessary until mixture has blended together.
4. Run blender at low to medium speed, then pour in melted oils in a thin stream. As the oils are blended in, the cream will turn white and the blender's motor will begin to grind.
5. When you have a mayonnaise-like consistency, stop the blender and add the essential oil.
6. Transfer the cream to glass jars while still warm because it thickens quickly.

Coconut butter body moisturizer

A very rich moisturizer good for dry or weather-beaten skin and sunburn.

You will need: 30ml (2 tbsp) beeswax, 10ml (2 tsp) distilled water, 110g (4oz) cocoa butter, 60ml (4 tbsp) sweet almond oil, 30ml (2 tbsp) coconut oil, double boiler, glass jar.

1. Melt the beeswax with the water over a low heat.
2. Remove from the heat, spoon in cocoa butter and blend.
3. Gradually blend in the oils.
4. Pour the mixture into a glass jar. The lotion will thicken as it cools.

Simple hand cream

A quick and easy-to-make hand cream that will soothe and protect your hands.

You will need: 55g (2oz) beeswax, 300ml (½ pint) sweet almond oil, 300ml (½ pint) water, 10 drops essential oil (if desired, for fragrance), double boiler, saucepan, blender.

1. Heat the beeswax and sweet almond oil in the double boiler until the wax melts.
2. In a saucepan, heat water until warm. Both mixtures should be warm, but not so hot as to be uncomfortable to the touch.
3. Place the warm water in the blender and

turn it on to high speed.

4. Slowly pour in the beeswax-oil mixture in a thin stream. When most of the oil has been added, the mixture should begin to thicken.

5. Add the essential oil if desired.

6. Turn off the blender when you have a thick cream.

7. Spoon into ointment jars or small metal tins.

Rich hand cream

A thick, enriched hand cream that will relieve and protect dry and chapped hands.

You will need: 225ml (8 floz) sweet almond oil, 100ml (3½ fl oz) coconut oil or cocoa butter, 5ml (1 tsp) lanolin, 15g (½ oz) grated beeswax, 200ml (7 floz) rosewater, 150ml (¼ pint) aloe vera gel, 1–2 drops rose blend essential oil, 5 drops vitamin E oil, double boiler, jar.

1. In the double boiler melt sweet almond oil, coconut oil or cocoa butter, lanolin and grated beeswax together over low heat and cool to room temperature.

2. When cool, add the other ingredients.

3. Whip the mixture to a smooth consistency.

4. Store in a covered jar.

Above *Beeswax has been used for centuries in a wide range of beauty products, most of which are easy to make at home.*

Recipes

Honey is nature's natural sweetener. It is more delicate, subtle and varied in taste than sugar. It has been used in cooking for centuries – the first known cookbook from ancient Rome used honey in all types of recipes, savoury as well as cakes and desserts. Honey goes particularly well with dairy produce such as cheese and yoghurt. The following recipes reflect its versatility in the kitchen, where it can as easily be used as a marinade for meat, a glaze for vegetables and a sauce for fruit. At the end of the summer you may have more honey than you know what to do with, so why not make some traditional honey cakes that can be stored throughout the year.

Roasted figs with Gorgonzola and honey

This is an exotic-looking and -tasting dish that makes quite an impression served as the first course at any dinner party. However, it is deceptively simple and quick to make. If you don't have Gorgonzola, any creamy blue cheese will work.

- 12 ripe figs
- 160g (6oz) Gorgonzola
- salt and pepper
- 30ml (2 tbsp) honey

Preparation time: 10 minutes

Cooking time: 8 minutes

Serves: 6

1. Pre-heat the grill to its highest setting.
2. Halve the figs and place them cut side up on an oiled baking tray.
3. Season with salt and black pepper and grill for 5–6 minutes, until they are soft and starting to bubble.
4. Chop the cheese into 1cm (½in) cubes.
5. Remove the baking tray from the grill and put the cheese on the top of each of the figs. Gently press the cheese down to squash it in a bit.
6. Grill for a further 2 minutes, until the cheese is bubbling and browning slightly.
7. Serve immediately with the honey poured over, and a few drops of balsamic vinegar for presentation.

- just under 2kg (4lb) spring lamb (leg or shoulder)
- 2 tbsp rosemary
- 1 tsp ginger
- 225g (8oz) honey
- salt and pepper

- 300ml (½pt) cider

Preparation time: 10 minutes

Cooking time: about 2 hours (allow 25 minutes per pound and 20 minutes extra)

Serves: 4–6

Honey-basted lamb

Cooked with cider and herbs, this is the perfect main course for Sunday lunch. Serve with garden potatoes and fresh mint sauce.

1. Preheat the oven to 225°C/430°F/gas mark 7.
2. Line an ovenproof dish with foil.
3. Prepare the lamb by rubbing with salt, pepper and ginger.
4. Place the lamb in the dish and sprinkle half the rosemary over it.
5. Pour honey over the top of the meat and cider around it, then put in the over.
6. After 30 minutes, reduce cooking temperature to 200°C/400°F/gas mark 6.
7. Baste during cooking, adding more cider if necessary.

Spicy chicken wing marinade

The aromas wafting across the garden as this marinated chicken sizzles on the barbecue on a warm summer's day are truly mouthwatering. It also has a sensational, rich taste as all the flavours have soaked into the meat.

1. The day before you want to serve this dish, whisk together the marinade ingredients.
2. Place the chicken wings in a large container, pour the marinade over them, cover and refrigerate overnight.
3. Place the marinated chicken on a hot barbecue and cook for 5–7 minutes, turning and basting frequently until cooked through but tender.
4. As an alternative to the barbecue, pre-heat the oven to 190°C/385°F/ gas mark 5 and bake the chicken wings in the marinade for 20 minutes or until cooked, turning and basting frequently.

- 60ml (4 tbsp) sesame oil
- 60ml (4 tbsp) white wine vinegar
- 45ml (3 tbsp) crunchy peanut butter
- 45ml (3 tbsp) clear honey
- 30ml (2 tbsp) chilli sauce
- 900g (2lb) chicken wings

Preparation time: 12 hours
Cooking time: 20 minutes
Serves: 4–6

Mustard and honey salad dressing

This popular sweet salad dressing is easy and quick to make. The hot mustard combines well with the sweetness of the honey and the sharpness of the lemon to create a rich-tasting dressing that goes with most salads.

- 5ml (1 tsp) clear honey
- 5ml (1 tsp) Dijon mustard
- 30ml (2 tbsp) lemon juice
- pinch of salt
- 60ml (4 tbsp) sunflower oil or a mixture of sunflower and olive oil

1. Mix the first four ingredients together.
2. Add oil and whisk together.
3. Pour into a jug.

Amber shallots

These deliciously sweet shallots go well as a side dish with meat dishes. They are easy to prepare but make sure you leave enough time to cook them so that they are really tender.

- 8 shallots
- 1 tsp salt
- ½ tsp paprika
- 30g (1oz) butter or margarine, melted
- 150ml (¼ pint)

tomato juice
- 45ml (3 tbsp) honey

Preparation time: 15 minutes
Cooking time: 40 minutes– 1 hour
Serves: 8

1. Peel the shallots and half any large ones.
2. Place in a large, shallow, greased casserole.
3. Combine salt, paprika, melted butter, tomato juice and honey.
4. Pour sauce over onions.
5. Cover dish and bake at 150°C/300°F/gas mark 3 for 1 hour or until the onions are tender.

Honey-glazed vegetables

Ideal with a roast dinner. The delicious honey juices combine beautifully with the meat, but also go perfectly with a nut roast. You can really use any vegetable you like, rather than just those listed here.

- 2 parsnips
- 2 carrots
- 2 red onions
- 30g (1oz) butter
- 30ml (2 tbsp) honey
- 30ml (2 tbsp) lemon juice
- 250ml (9 floz) vegetable stock
- salt and pepper

Preparation time: 5 minutes
Cooking time: 30 minutes
Serves: 4

1. Pre-heat the oven to 200°C/400°F/gas mark 6.
2. Peel and prepare the vegetables. Cut in half lengthways, then slice into 1cm (½in) pieces.
3. In a large flat pan, gently soften the onion and then remove.
4. Combine carrots, butter, stock and a pinch of salt. Bring to the boil, lower the heat to medium, and cover and simmer for 4 minutes.
5. Add the parsnips and cook for about 5 minutes or until the vegetables are tender. Now return the onion to the pan, and transfer the vegetables to a shallow baking dish with a draining spoon, preserving the liquid in the pan. Add salt and pepper.
6. Make sure you have 250ml (9 floz) of stock remianing (adding more if necessary). Stir in the honey and lemon juice. Bring to the boil and simmer for 1 minute or until the liquid has thickened slightly. Pour over the vegetables.
6. Cook uncovered for 15–20 minutes or until lightly glazed.

Traditional honey cake

Honey cake is the oldest cake. It was made in biblical times and there are pictures of Egyptians baking it in pharaohs' tombs. It became spice cake and gingerbread in the Middle Ages and is still enjoyed today. It was often made at the end of the summer when the honey and grain were harvested and is still traditionally served at this time as part of the Jewish New Year celebrations to wish for a sweet year ahead. It is a cake to accompany a cup of tea, rather than a dessert after a meal.

- 400g (14oz) plain flour
- 175g (6oz) firmly packed dark brown sugar
- 1 tbsp baking powder
- 1 tsp baking soda
- 2 tsp cinnamon
- ½ tsp each ground allspice, nutmeg, ginger and salt
- 4 eggs

- 450ml (1 pint) honey
- 60ml (4 tbsp) vegetable oil or melted butter
- 200g (7oz) coarsely chopped nuts (optional)

Preparation time: 30 minutes
Cooking time: 1 hour
Serves: 10–12

1. Preheat the oven to 175°C/350°F/gas mark 4.
2. Grease and flour two 22 x 12cm (9 x 5in) loaf tins.
3. Stir all dry ingredients together until well combined.
4. Make a well in the centre of the mixture and add eggs, honey and oil.
5. Stir mixture until smooth.

6. Divide the mixture equally between the prepared tins.
7. Bake in the oven for about 1 hour or until a toothpick inserted into the middle comes out clean.
8. Cool for 10 minutes before removing the cakes from the tin.
9. Leave the cake to cool further on a wire rack.
10. Sandwich the two layers together with honey and allow to mature for a day or two before eating.

Honeyed yoghurt breakfast

There is no better breakfast than honey dripped over rich yoghurt. It is nourishing and delicious as the sweetness of the honey offsets the sharp tang of the yoghurt. Add a slice of chopped melon or a chopped-up banana for your morning portion of fruit.

- 150ml (½ pint) thick Greek yoghurt
- 15g (½oz) toasted flaked almonds
- 25g (1oz) raisins
- 1 banana or slice of melon chopped

- 30ml (2 tbsp) clear honey

Preparation time: 2 minutes
Serves: 4

1. Mix together yoghurt, almonds and raisins in a bowl.
2. Chop up a banana or pieces of melon and add to the mixture.
3. Transfer to a glass serving bowl and drizzle with honey and serve.

Mead

Dating back more than 3,000 years, mead has been enjoyed by the great civilizations of the world. It is also called honey wine because its alcohol content is similar. It tastes like sherry and, like wine, it matures well. Your guests will be very impressed if you serve it at a party – although it can be an acquired taste!

You will need two large fermenting vessels with airlocks, and a siphon or funnel, all of which you can buy from a wine-making supplier.

- 1.4kg (3lb) honey
- 4 litres (7 pints) water – preferably filtered rainwater
- 1 package wine yeast

Preparation time: 1 month
Cooking time: 1 year
Makes: 4.5 litres (1 gallon)

1. Heat about half the water to boiling point in a large, clean, stainless steel pot.
2. Take the water off the boil, and add about two-thirds of the honey, reserving 450g (1lb) for later sweetening. Stir until the honey dissolves. Add the remaining water. Heat again and remove any scum that may appear. Cool this must (the mixture of honey and water) to around 27°C (80°F).
3. Add the yeast and agitate vigorously to mix it in and to give the must lots of oxygen.
4. Pour this mixture into your fermenting vessel.
5. Install the airlock, place the mead in an area that will maintain a more or less constant temperature of 21–27°C (70–80°F) and wait, making sure the airlock has liquid in it at all times. You should notice the airlock start to bubble in a day or so and continue to do so for about a month.

6. After about a month when the bubbling has stopped, siphon the must into a clean fermenting vessel and airlock again. At this stage taste the liquid and add sterilized honey water if you would like to sweeten the mead. (To make this, stir 1 part honey into 2 parts boiled water and let it cool before adding it to the mead via a siphon or funnel.)
7. Watch your airlock, and when the bubbling has slowed to less than 1–2 bubbles per minute, you can either store the mead for bulk aging or siphon it into sterilized bottles.

Beer bottles with caps are best since they can withstand any pressure that may build up if the mead continues to ferment. Wine bottles with corks can be used if the fermenting has truly finished.
8. Leave for a year before tasting.

You can vary the flavour of mead by adding herbs and spices to the recipe. Cloves, cinnamon, marjoram and oranges are just a few of the flavours that can be added.

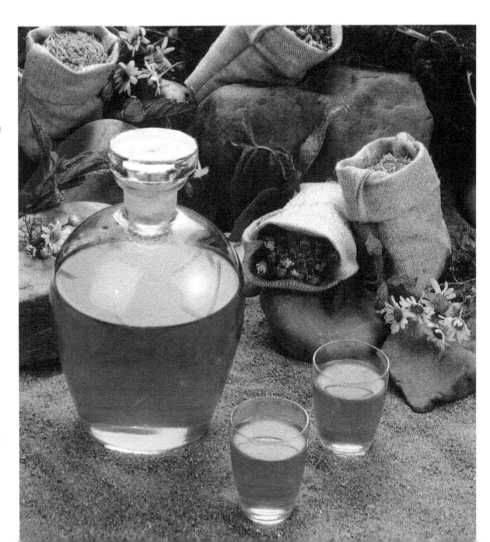

Health and care

Good hygiene is essential for the health of your bees and is easily achieved by keeping your apiary clean and tidy and not interchanging equipment and components between hives. Sometimes, however, ailments can occur in even the cleanest apiary, since bees are susceptible to a number of parasites and diseases. By taking sensible precautions against the most common pests and following an integrated pest-management system you should be able to maintain strong, vigorous colonies that can use their own defences to ward off foes.

Prevention and cure

It is unfortunately a fact of nature that honey bees are susceptible to a number of parasites and diseases both in the wild and in the hive. As a beekeeper you will come across the common pests and parasites, such as wax moth and varroa, but hopefully the more serious foulbrood disease will be something that you only read about.

There are a number of different approaches to helping bees deal with their maladies but the most important is taking preventative action to keep your colonies strong. Strong colonies use their own natural defences to combat their adversaries and diminish the need for you to act as a doctor.

Below *Keeping your apiary clean and tidy and ensuring your colonies are strong will help the bees fight disease.*

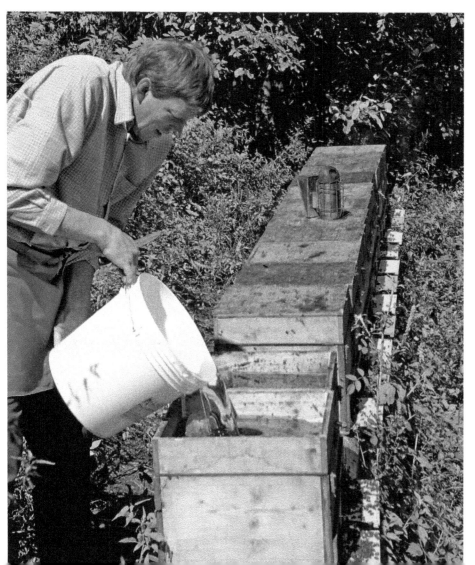

Preventative measures

- Keep your apiary clean and tidy.
- Only buy bees that are from disease-free, reputable apiaries.
- Dispose of old comb and propolis rather than leaving it around the hive.
- Don't buy second hand frames since they can't be effectively sterilized.
- Sterilize by scorching second hand hives – or avoid buying.
- Sterilize hive parts and equipment if moving between apiaries.
- Combine weak colonies. One strong colony is better than two weak ones.
- Try to prevent bees robbing (stealing honey from another hive) and drifting (moving into a neighbouring hive) by keeping your entrances small enough to defend. If you have a number of hives, face the entrances in different ways so that the bees know which hive is theirs.
- Keep up-to-date with any new methods of pest control and follow an integrated pest management (IPM) routine throughout the year.
- Keep good records so you can monitor changes in the hive and know what medication has been administered.
- If your bees die, close the hive to prevent other bees getting access to the honey and picking up the disease.

Warning signs

Unlike with pets, it is hard to tell if your bees are off colour by their behaviour. With experience you will be able to spot some changes by looking at the adult bees but the

Right *A healthy-looking brood box with lots of activity means that your bees are doing well.*

INTEGRATED PEST MANAGEMENT

The term 'integrated pest management' (IPM) has become more common in beekeeping since the varroa mite became so widespread. Instead of depending on man-made chemicals to fight parasites and disease, IPM is a more holistic approach using a combination of techniques that may include natural methods to enable the bees' own defence mechanisms to create a strong colony. One example is using an open mesh floor where the varroa mite, which has been preened off an adult bee, drops through the hive onto the ground where it can't climb back up into the hive. Your local beekeeping association will have more details of IPM techniques and will run courses on IPM systems that are recommended by the national body that oversees animal husbandry where you live.

most reliable way to assess the health of a colony is to check what is going on in the brood. A healthy looking brood has a consistent pattern of brood laying pearly white larvae where the segments are clearly visible, and uniform brood cappings of a light brown colour. There should be no unpleasant smells and the hive will be free from damp and mould. There should be stores of fresh-looking pollen and white-capped honey cells and your adult bees should be lively with perfect wings and appear to be flying well. If, during your regular hive inspections, the colony isn't performing as above, you should seek assistance from an experienced beekeeper who will be able to identify the problem and advise on a course of action.

There are a number of ailments that affect the bee brood which are normally noticeable because the larvae are not looking pearly-white, segmented and C-shaped. In the UK two diseases – American Foulbrood and European Foulbrood – and one pest – small hive beetle – are notifiable, which means that

if you suspect an outbreak of any of these you must, by law, inform the National Bee Unit (see 'Getting your bees'). They will send out an inspector or ask you to submit a suspect disease sample for analysis.

> *IPM takes a more holistic approach, using more natural methods to enable the bees' own defence mechanisms to create a strong colony.*

Brood diseases

American Foulbrood (AFB) and European Foulbrood (EFB)

The geographical names of these two rare but serious diseases caused by bacterium derive from where they were discovered, but somewhat confusingly they can occur in any country throughout the world. If your brood is not looking pearly white with the older larvae curving around into a C-shape within the cell, this is the first signs that something is wrong.

If the capped cells are sunken and dark in colour, suspect AFB. The test you can administer is called the 'ropiness' test. Insert a matchstick into the sunken cell and twist it. If, when you pull it out of the cell, there is a long, unbroken, mucus-like thread attached, then your colony has AFB. There may also, but not always, be an unpleasant smell. Call your regional bee inspector or local association immediately for help. If your analysis is correct the recommended action is for your frames and the bees to be destroyed, by burning, to prevent it spreading to other colonies.

With EFB, you will see larvae that look as if they have melted and are turning a yellowish-brown colour. EFB is harder to diagnose but there are kits available for you to use. If you suspect EFB it is best to call an experienced beekeeper for their advice. EFB can be treated if it is a small infection.

Small hive beetle

The small hive beetle (SHB) originates in Africa where the native honey bees *Apis mellifera capensis* and *Apis mellifera scutellata* can deal effectively with their natural predator. The European honey bee, however, can't cope with this tropical pest which has wreaked havoc in apiaries across the south-eastern states of the USA. The dark brown, nearly black, oval beetle is about 7mm (¼in) long and 4mm (⅛in) wide and lives and breeds in the colony of honey bees. Females lay masses of eggs. The larvae feed on the bees' brood, pollen and honey.

The mass of larvae often congregate together but they don't produce silk threads that look like webbing (it's the larvae of the common wax moth – see below – that do that). When you open an infected colony you will see the beetle scarper to the dark corners of the hive and its white larvae will be all over the brood, eating it. If you suspect your colony has been infected by SHB contact your regional bee inspector or your local association immediately. These beetles probably enter the country by container so if you are located near points of entry such as major airports or ports be extra vigilant.

Chalkbrood

A much more common but less serious brood disease is chalkbrood, which is caused by a fungus which germinates in a stressed colony. This can occur when the colony is stressed possibly due to damp conditions. The fungus grows and is ingested by the larvae, which die shrunken and hard. The chalk-white remains can easily be seen on the hive's landing board as the house bees try to clean out the dead larvae. A strong colony can live with a small amount of chalkbrood. Cleaning out the brood frames and replacing the existing queen with a healthier one will help if the problem continues.

Adult diseases and parasites

Nosema

There are two types of nosema; *nosema apis* and *nosema cerana*. Both are caused by a parasitic fungus which lives in the gut of the adult bee. It impairs the digestion of pollen and gives the bee stomach ache. This illness is passed to the rest of the colony via the faeces. Nosema can only be diagnosed by examination under a microscope but symptoms include diarrhoea – dark-coloured excrement – on the outside of the hive and possibly on the frames. It weakens the bees and reduces their life expectancy so the colony becomes depleted.

Nosema cerana is endemic in the eastern honey bee, but has recently spread to our western honey bee, which has not had time to build up natural defences against it. It is the more vigorous variety and can lead to a lack of vigour in the bees and cause heavier intestinal injuries. It has been implicated in the death of bees from colony collapse disorder.

Going into winter with a healthy, vigorous colony with good honey stores will help to reduce the likelihood of Nosema. Replacing old brood frames in the spring should also reduce the likelihood of the parasite as any infected comb is destroyed.

Tracheal mites (Acarine)

The mite *Acarapis woodi* lives in the breathing tubes of the adult bee and causes premature death, usually in winter, leading to 'spring dwindling' in the colony. The expanding brood is left unattended by the drop in the adult population and the colony is severely weakened. Only diagnosed by microscopic examination, the symptoms can be deformed wings and weak bees with distended bodies. There is no effective treatment as yet. Keeping strong, vigorous colonies and good husbandry is recommended. There is evidence that the chemicals used for varroa are also having an effect on this mite. Keep up-to-date with current treatments by talking with your local bee association.

Going into winter with a healthy, vigorous colony with good honey stores will help to reduce the likelihood of Nosema.

tip
If your bees die, close the hive to prevent other bees getting access to the honey and picking up the disease.

Varroa

Varroa (*Varroa destructor*) is a mite that lives on both the brood and the adult bees. If left untreated your colony will perish. It has become a worldwide problem and varroa is becoming resistant to some of the treatments available to the beekeeper.

The reddish brown mite, about 1.5mm (¹⁄₂₀in) long, lives and breeds within the honey bee colony and in the process debilitates and kills the colony.

Varroa arrived from Asia where the *Varroa jacobsoni* lives on the Asian honey bee species, a honey bee that can co-exist with the parasite. The varroa is thought to have developed a new species – the aptly named destructor – designed to attack the western honey bee. Since our honey bee has not had an evolutionary history of millions of years with the mite it has not been able to develop any defence mechanisms to combat varroa.

The mite is most easily seen on the white larvae. But it can also be seen if you look carefully on the adult bees' thorax and also

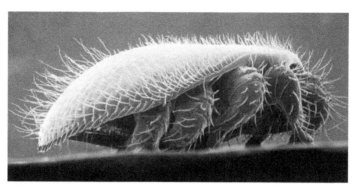

Left *Varroa is a 1.5mm long mite which debilitates and kills honey bee colonies.*

running around the comb. The mite's home is solely within the bees' nest, feeding on the brood and it breeds on the pupae within the capped cells.

Varroa life cycle

The female mite will arrive at a new bee colony by hitching a lift on the back of a foraging bee or a drone. The mite doesn't fly but can jump from one bee to another remarkably quickly. So if bees meet at a flower the varroa can jump across to another bee and find its way into a new colony, or bees can occasionally drift from one colony to another and will transport the mite that way.

Once there, the female mite will look for a fat larvae in a cell that is about to be capped over; there she lays her eggs. When the varroa eggs hatch they feed on the pupating bee. The varroa can lay 4–5 eggs in one cell, one male and the rest females, and when they mature (all still within the capped cell) they mate. When the cell is uncapped the varroa emerge. The male mite will die but the mated females survive and will find a new fat bee larvae and repeat the process. Only this time there are 3–4 females all laying 4–5 eggs

going on to produce 20 new females which will produce 80 more females, and so on. It is this rapid explosion of the varroa population within a colony that causes it to decline.

As well as damaging the pupae by sucking its blood the varroa spreads many damaging viruses that can spread quickly around a colony. These lethal viruses destroy the balance of the colony by reducing the life span of the adult bees. This imbalance will eventually cause the colony to die. For example, infected foragers will prematurely die so the colony will not receive enough food to maintain the brood and the colony will collapse. The viruses affect the bees in many ways and we can only see the symptoms of a few of them. Deform wing virus is one that we can see quite easily. The misshapen and dishevelled wings of an adult bee are the obvious signs and means that the virus has taken hold.

One of the jobs of a beekeeper is to keep the varroa population to a minimum. A strong colony can live with a small number of varroa so our aim is to keep the mite population as low as possible.

Below *The parasitic varroa mite is most easily seen on the white bee larvae. It feeds on the brood.*

Control methods

We need to implement a series of measures, which we call Integrated Pest Management (IPM), that when combined will help to control the varroa population. This IPM system takes a holistic approach that includes using man-made chemicals to help fight the parasite but also more natural methods that enable the bees' own defence mechanisms to create a strong colony. These are sometimes referred to as hard and soft techniques or chemical or biotechnical methods.

Whichever treatment you decide to use, the first step is to monitor the levels of varroa within your colony. There are two main ways to monitor varroa levels. One is to use an open mesh floor on your hive with a monitoring tray inserted below, and the other is to take a sample of drone brood and count the number of cells containing varroa.

Open mesh floors

Open mesh floors use a mesh screen on the floor of the hive which mites fall through onto a monitoring tray below. You can then count the tiny mites on the tray. They can be differentiated from other debris, such as pollen and wax flakes, because they are shiny, oval and reddish-brown. Monitor the mite drop over a number of days and work out the daily mite drop by dividing the number of mites counted by the number of days. It is best to do this over a 14-day period since this will give you a more accurate figure. It may be easier to check the tray every 3–4 days and count the mites, keep a running total and clean the tray after each count. At the end of the 14 days you will have your total.

The number of mites you find will depend on the time of year you monitor but a daily

Above **Bees on a frame with brood, pollen and nectar but to the right of the bees is a small, brown varroa mite.**

drop of over 20 mites in the summer and over 6 mites a day in the spring means your colony is in trouble and you should take action. There are varroa calculators on the internet for more accurate diagnosis.

Brood sampling

Female varroa prefer to lay their eggs in the cells of drone brood since the drone pupae is capped for 24 days, 3 days longer than the worker bees, which gives the mites more time to lay, hatch, mature and mate in the capped cell. So we select drone brood for monitoring.

To encourage your bees to make drone brood for monitoring purposes you can

replace a deep frame in your brood box with a shallow frame of foundation. The bees will add wild comb, normally in the form of drone cells, to the bottom of the shallow frames. This will be used for the varroa monitoring.

> To encourage your bees to make drone brood for monitoring purposes you can replace a deep frame in your brood box with a shallow frame of foundation.

When the drone brood is capped (the varroa will be trapped in the cells), remove it and using an uncapping fork dig out 100 drone pupae. Count the number of cells infested with varroa. If more than 10 per cent of the drone cells you have uncapped are infested then you need to take action. Sacrificing drone brood is not one of the nicest jobs a beekeeper has to perform but it could help the health of your colony.

Varroa treatments

There are a number of ways to treat against an infestation of varroa and the purpose of all of them is to reduce the number of varroa within a colony.

Treatments that use varroacides to destroy the varroa by means of adding a chemical need to be used with caution. We don't want the chemicals to get into the food chain through the honey we harvest, so take advice from your bee inspectors or local associations to find out what time of year they can be used and if they affect the honey. Some are best used after the final honey harvest of the year, others when the queen is not laying eggs for a short period in the mid winter.

Chemical treatments

- Pyretheroids: These chemical treatments usually come on impregnated plastic strips that are inserted between a couple of the frames. The varroa mite has developed resistance to these pyretheroids and so they may not be effective in your area. You can find out from your local supplier. Read the instructions carefully before administering the treatment.
- Thymol: At the end of the honey harvest a thymol-based treatment is currently recommended. The strong thymol fumes kill the varroa but only the varroa that are outside of the capped cells. The treatment needs to be carried out for at least four weeks for it to work effectively. The application also needs to be carried out when the temperature is above 15°C. Since the smell can get into the honey you need to take your honey off before starting the thymol treatment. There are a number of thymol products available to buy.
- Oxalic acid: If you do need to treat in the winter then you could use an oxalic acid treatment but this should only be used when there is no brood as the varroa will be on the adult bees and will die on contact with the acid. Not all colonies go broodless so this treatment is less effective if your colony has brood in mid-winter.
To ensure you administer the right dose, buy a ready-prepared solution of the acid in sugar syrup. It comes in a special bottle which allows you to trickle just 5ml (1tsp) on each seam – the space between the frames – of live bees. There could be 4–8 seams of live bees in mid-winter. You will need to open the hive on a cold day and quickly administer the warmed solution. Although this treatment is applied in mid-winter when your bees should be very quiet, you should still wear your bee suit and gloves for protection.

Non-chemical treatments

- Sacrificing drone brood: Since we know that varroa prefer to breed in drone cells we can strategically raise some drone brood to trap the varroa. Once the cells are capped and the varroa are inside we cull this extra drone brood which will remove a percentage of the varroa. This can be an effective control but don't cull all the drones since they are needed for the next

Above *Dousing your bees with icing sugar is part of varroa control.*

generation. You can raise drone brood by adding a shallow frame into the brood box. The bees will more than likely make drone-sized cells (birthing chambers) hanging from the bottom of the shallow frame. You could also cut out a section of foundation from a brood frame and the bees will replace it with drone sized cells.
- Sugar dusting: Bees are continually keeping each other clean by preening and can occasionally pick off a varroa from their sister's thorax but this can be accelerated by dusting the bees with castor sugar to encourage the preening. Use a sieve and lightly dust each frame of bees with the sugar. This needs to be done in conjunction with an open mesh floor so the mites drop to the ground and cannot climb back up into the hive. This method of varroa control is not effective by itself so it needs to be done in conjunction with other treatments.

Pests
Wax moth

Wax moth *(Galleria mellonella)* and their larvae cause a nuisance but can be kept under control by the bees and the beekeeper. The moth lays eggs in the hive and as they grow the caterpillars feed on the wax, pollen, honey and larvae. In a strong colony the bees' own police force will deal with this threat by chasing the moth and the larvae out of the hive, and you can easily see and remove them too. However, you won't be able to see the eggs, so in a weaker colony the moth may take over and eventually the colony will vacate the hive looking for a new home.

Moths' eggs can stay in the hive's boxes and frames even when they are not being used but they are susceptible to the cold (less than 2°C/36°F) so you can kill them by leaving hive components that aren't being used outside over the winter. If you live somewhere where the temperature doesn't fall this low, put the components in a freezer instead. Don't leave old comb and wax around the apiary, since the wax moth will surely find it and take hold.

- Mesh Floors: Using open mesh floors throughout the season (without the monitoring tray) allows the live mites to fall to the ground and they won't be able to climb up and return to the colony. The open mesh floor method alone is not sufficient in combating varroa so it needs to be used with other methods.

There are a number of more advanced non-chemical techniques to control varroa such as artificial swarm methods and queen trapping, but they are not recommended for the beginner beekeeper. Once you are more confident in handling your colony ask for advice from an experienced beekeeper who can explain these methods.

Varroa control is a crucial element of modern beekeeping that will require your attention, dedication and continued vigilance. Scientists are trying to develop a varroa-resistant western honey bee but until this happens, if it ever does, you should keep abreast of how best to deal with the problem as innovative methods enter the market every season.

Predators

Bears

In some parts of the world bears can be a problem for your apiary. They like to eat bee brood as well as the honey and will cause much damage to the hives in the process. If you live in an area with bears keep your hives away from its paths, ravines and forests.

Skunks

Strong colonies can overwhelm this insectivore but weaker colonies might lose out. Protection by fencing and stands can prevent the skunk from scratching at the entrance of the hive. When the bees come out to defend their home they fall prey to the hungry animal.

Badgers

Badgers are keen on honey too. They tend to push hives over by getting underneath them and doing a press-up with their powerful front legs which then topples the hive. Many common hive stands are the perfect height for badgers, so if you are in a badger area you'll need to put up a barrier to stop them getting to the hive.

Woodpeckers

Woodpeckers love to peck away at the side of a wooden hive to get to the grubs and bees inside which they feed on. As well as depleting the colony size, the holes created by the woodpecker can also cause ventilation problems in the hive. This pest is not confined to the rural beekeeper as the red-tailed greater spotted woodpecker is becoming more common in the urban environment.

To deter the bird you will need to surround your hive with a wire mesh screen. This could be on a frame structure or just circled around the hive. It needs to be far enough away from the hive so the bird doesn't use it as a perch to hang on while it happily pecks away at the hive.

There are a number of more advanced non-chemical techniques to control varroa such as queen trapping, but they are not recommended for the beginner beekeeper.

Above *A weak honey bee colony that has been infested with wax moth and their larvae.*

Colony Collapse Disorder

Colony Collapse Disorder (CCD) is the name that has been given to the mysterious death of billions of honey bees across the globe since 2007. It was first reported in the USA among commercial beekeepers who described how adult bees had vanished from the hives and were leaving behind just the queen and a patch of brood. This very unusual behaviour captured the public's imagination and raised awareness about honey bees worldwide, with apocalyptic headlines about the end of the world being nigh. It has been perplexing the beekeeping world ever since.

Though the problem was mainly found in the USA, large-scale bee losses were seen in parts of Europe, especially in France and Germany. No single cause for this strange phenomenon has been found. Initially a virus was blamed; it shortens the bees' lifespan so adult, forager bees die on the wing outside the hive days before the brood has developed into new adult bees, leaving the colony with just a queen and brood and no bees to bring in food, feed the larvae and do all the other duties that are necessary in a functioning hive. As a result the colony dwindles and dies. But it transpired that the virus had been around for a while. The question was why had the bees become so susceptible to it now?

Scientists have now attributed the puzzling phenomenon to a combination of three main factors. Parasites – such as the ubiquitous varroa mite (see 'Varroa'), which spreads the viruses – a fungus that causes a bee gut disease, which also reduces bees' life expectancy and causes a colony to become depleted and die out – as well as pesticides and poor nutrition.

A new class of pesticide has been found in laboratory studies to interfere with bees' communication and mapping systems making it impossible to find their way home. It is toxic to bees and works its way through a plant from the seed, through the stem and leaves and into the pollen and nectar of the flower. In

Below *All these stacked beehive boxes held bees that mysteriously disappeared. Commercial bee farmer, Brumley's Bees, in California, lost 2,000 hives out of their 6,000 honey bee colonies when Colony Collapse Disorder was at its height.*

CCD is a problem to be aware of, but for hobby beekeepers it is rarely encountered.

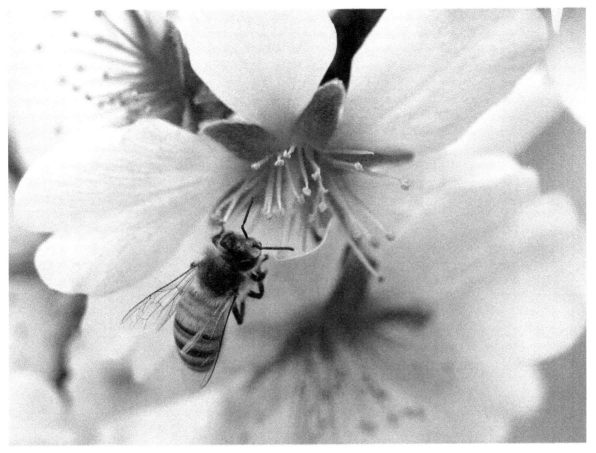

France, some of these systemic pesticides have been banned for use on certain commercial crops for allegedly killing bees. Following large bee losses in Italy, Germany and Slovenia, more were temporarily suspended pending further studies. There are concerns that even at minuscule quantities these neonicotinoid pesticides could have a detrimental impact on the health of the bee colony, or that they could synergize with other chemicals used in the field and significantly increase their toxicity to bees. Since the bee toxicity of licensed 'crop protection products' is only tested on adult bees, their potential impact on the workings of the whole colony are still unknown.

Bees need a diverse diet provided by the amino acids in pollen from a variety of trees, plants and flowers to thrive. But modern day agricultural practices create hundreds, if not thousands, of acres of just one crop for bees to eat. Moreover, certain monocrops that commercial beekeepers take their bees to pollinate, such as sunflowers and blueberries, are of particularly poor nutritional value, which can cause bees to become malnourished and weak.

CCD is a problem to be aware of, but for the hobbyist, it is rarely encountered. Good husbandry, varroa control, and keeping your bees strong and healthy are the best ways to combat any of the bees' many assailants.

Glossary

Abdomen
The posterior or third region of the body of the bee that encloses the honey stomach, stomach, intestines, sting and reproductive organs.

Acarapis woodi
A mite, also called the tracheal mite, which infests the bees' breathing or tracheal system.

American foulbrood (AFB)
A **brood disease** of honey bees caused by the spore-forming bacterium, *Bacillus larvae*.

Anaphylactic shock
Constriction of the muscles surrounding the bronchial tubes of a human, caused by hypersensitivity to venom (including bee stings). Requires immediate medical attention.

Apiary
The location and total number of hives and other beekeeping equipment at one site.

Apiculture
The science and art of raising honey bees.

Apis mellifera
A native European bee that is kept for its honey and wax in most parts of the world. Commonly known as the European or western honey bee.

Bacillus larvae
The bacterium that causes **American foulbrood.**

Bee brush
A soft brush or whisk (or handful of grass) used to remove bees from frames.

Bee escape
A device constructed to permit bees to pass one way within a hive, but prevent their return; used to clear bees from **supers**.

Beehive
A box or receptacle with movable frames, used for housing a colony of bees.

Bee space
A space big enough to permit free passage for a bee, too small to encourage comb building and too large to induce the bees to produce propolis; for *Apis mellifera* it measures 9.5mm (just over ⅓in).

Bee suit
Protective coveralls to protect from stings and to keep clothes clean.

Bee veil
A cloth or wire netting for protecting the beekeeper's head and neck from stings.

Beeswax
A waxy substance secreted by bees through special glands.

Black scale
The 'dried down' appearance of a larva or pupa which has died of a **brood disease**.

Brace comb
A bit of comb built between two combs to fasten them together, between a comb and adjacent wood, or between two wooden parts such as top bars.

Brood
Immature stages of bees not yet emerged from their cells; the stages are **egg, larva** and **pupa**.

Brood chamber
The part of the hive in which the **brood** is reared; may include one or more hive bodies and the combs within.

Brood diseases
Diseases that affect only the immature stages of bees, such as **American** or **European foulbrood.**

Brood nest
The part of the hive interior in which brood is reared.

Burr comb
Small pieces of comb made as connecting links between combs or between a frame and the hive itself.

Cage shipping
Also called a package, a screened box filled with 1–2.25kg (2–5lb) of bees, with or without a queen, and supplied with a feeder can; used to start a new colony, or to boost a weak one.

Capped brood
Immature bees whose cells have been sealed over with a brown wax cover by other worker bees.

Cappings
The name given to the thin wax covering over honey once it has been cut off the extracting frames; a source of premium beeswax.

Carniolan bees
A greyish race of honey bee, *Apis mellifera carnica.*

Cast
A second or subsequent **swarm** which leaves a colony with a virgin queen, after the first (or prime) swarm has departed.

Castes
The three types of bees that comprise the adult population of a honey bee colony: **workers, drones** and **queen**.

Cell
A single hexagonal compartment in a honeycomb.

Chalkbrood
A disease affecting bee larvae, caused by a fungus *Ascosphaera apis*. The larvae eventually turn into hard, chalky white 'mummies'.

Chilled brood
Immature bees that have died from exposure to cold; commonly caused by mismanagement.

Cluster
A large group of bees hanging together, one upon another.

Cocoon
A thin silk covering secreted by larval honey bees in their cells in preparation for pupation.

Colony
The aggregate of worker bees, drones, queen and developing brood living together as a unit in a hive or other dwelling.

Comb
The wax portion of a colony in which eggs are laid and honey and pollen are stored.

Comb honey
Honey in the wax combs, usually produced and sold separately.

Creamed honey
Honey that has been pasteurized and undergone controlled granulation to produce a finely textured candied honey which spreads easily.

Cross-pollination
The transfer of pollen from the anther of one flower to the stigma of another of the same species.

Dividing
Separating a colony to form two or more units.

Drawn comb
Combs with cells built out by honey bees from a sheet of foundation.

Drifting
The movement of bees that have lost their location and enter other hives.

Drone
The male honey bee which comes from an unfertilized egg.

Drone brood or drone comb
Brood which matures into drones, reared in cells larger than those of worker brood.

Drone congregating area (DCA)
A specific area to which drones fly waiting for virgin queens to pass by; it is not known how or when they are formed, but drones return to the same spots year after year.

Drone-laying queen
A queen who can lay only unfertilized eggs, due to age, improper or no mating, disease or injury.

Dwindling
The rapid dying-off of old bees in the spring; sometimes called spring dwindling or disappearing disease.

Dysentery
An abnormal condition of adult bees characterized by severe diarrhoea and usually caused by starvation, low-quality food and/or moist surroundings.

Eggs
The first phase in the bee life cycle enclosed in a flexible shell or chorion.

Entrance block
A notched wooden strip used to regulate the size of the entrance to a hive.

European foulbrood (EFB)
An infectious **brood disease** of honey bees caused by the bacterium *Streptococcus pluton*.

European honey bee
See *Apis mellifera*.

Extractor
A centrifugal force machine to throw out honey but leave the combs intact. May be either radial (working on both sides at once) or tangential (you have to turn it).

Feeders
Various types of appliances for feeding bees artificially.

Fermenting honey
Honey which contains too much water (greater than 19 per cent) in which a chemical breakdown of the sugars takes place, producing carbon dioxide and alcohol.

Field bees
Another name for **foragers**.

Flight path
Usually refers to the direction in which bees fly when leaving their colony; if the path is obstructed, the bees may become aggravated.

Forage
Natural food source of bees (nectar and pollen) from flowers.

Foragers
Worker bees which are usually 21 or more days old and work outside to collect nectar, pollen, water and propolis.

Foundation, wax
Thin sheets of beeswax embossed with the base of a worker cell on which bees will construct a complete comb (**drawn comb**).

Foundation, wired
Comb foundation which includes wires for added support.

Frame
Four pieces of wood forming a rectangle designed to hold honeycomb.

Fume board
A rectangular frame, the size of a super, covered with an absorbent material such as burlap, on which is placed a chemical repellent to drive the bees out of supers for honey removal.

Guard bees
Worker bees about three weeks old who guard the entrance to the hive.

Hive
A manmade home for bees.

Hive body
A wooden box containing frames.

Hive stand
A structure serving as a base support for a beehive; it extends the life of the bottom board by keeping it off damp ground.

Hive tool
A flat metal device with a curved scraping surface at one end and a flat blade at the other; used to open hives and to prise apart and scrape frames.

Honey
A sweet viscid material produced by bees from the nectar of flowers, composed largely of a mixture of dextrose and laevoluse dissolved in about 19 per cent water; contains small amounts of sucrose, mineral matter, vitamins, proteins and enzymes.

Honeydew
An excreted material from insects in the order Homoptera (aphids), which feed on plant sap.

Honey colour
Measured by a Pfund grader, honey colours are classified into seven gradations from water white and white to amber and dark amber.

Honey flow
A time when nectar-bearing plants are blooming, allowing bees to store a surplus of honey.

Honey sac
The part of the body in which bees carry nectar.

Honey supers
Hive bodies used for honey production.

Inner cover
An insulating cover fitting on top of the top super but underneath the outer cover, with an oblong hole in the centre.

Inverts
An enzyme in honey, which splits the sucrose molecule (a

disaccharide) into its two components dextrose and laevulose (monosaccharides).

Italian bees
A common race of honey bees, *Apis mellifera ligustica*.

Larva, capped
The second developmental stage of a bee, ready to pupate or spin its cocoon (about ten days after its emergence from the egg).

Laying workers
Worker bees which lay eggs in a colony that is hopelessly queenless; since the workers cannot mate such eggs are infertile and therefore produce drones.

Mandibles
The jaws of an insect; used by bees to form honeycomb, to scrape pollen, in fighting and in picking up hive debris.

Mating flight
The flight taken by a virgin queen during which she mates in the air with several drones.

Mead
Honey wine.

Metamorphosis
The three stages through which a bee passes before reaching maturity: **egg, larva** and **pupa.**

Migratory beekeeping
The moving of colonies of bees from one locality to another during a single season to take advantage of two or more **honey flows**.

Moisture content
In honey, the percentage of water should be no more than 18.6; anything higher than that will allow honey to ferment.

Movable frames
A frame constructed in such a way as to preserve the **bee space** and to be easily removed without damaging the bees; when in place, it remains unattached to its surroundings.

Natural honey
Unfiltered and unheated honey.

Nectar
A liquid rich in sugars, manufactured by plants and secreted by nectar glands in or near flowers; the raw material for honey.

Nectar glands
Special nectar-secreting glands usually found in flowers, whose function is to attract pollinating insects such as honey bees for the purpose of cross-pollination, by offering a carbohydrate-rich food.

Nosema disease
A widespread adult bee disease

caused by a one-celled, spore-forming organism, *Nosema apis*; it infects the bee's gut lining.

Nuc, nuclei, nucleus
A small colony of bees often used in queen-rearing or as a 'starter' hive.

Nurse bees
Young bees, three to ten days old, which feed and take care of developing **brood**.

Outer cover
The last cover that fits over a hive to protect it from rain; the two most common kinds are telescoping and migratory covers.

Package bees
See **Cage shipping**.

Pheromone
A substance secreted by an animal which causes a specific reaction, such as stimulation to mate with or supply food to an individual of the same species.

Piping
A series of sounds made by a queen, frequently before she emerges from her cell.

Play flights
Short flights taken in front and in the vicinity of the hive by young bees to help them familiarize themselves with its location.

Poison sac
Large oval sac containing venom and attached to the anterior (front) part of the **sting**; stores venom produced by the poison gland; its primary ingredients are peptide and mellitin.

Pollen
The dust-like male reproductive cells (gametophytes) of flowers.

Pollination
The transfer of pollen from the anthers to the stigma of flowers.

Prime swarm
The first **swarm** to leave the parent colony, usually with the old queen.

Proboscis
The mouthparts of the bee that form the sucking tube or tongue.

Propolis
Plant resins collected and modified by bees, used to fill in small spaces inside the hive.

Pupa
The third stage in the development of the bee during which it is inactive and sealed in its cocoon.

Queen
A fully developed and mated female bee, responsible for all the egg-laying of a colony and recognized by other bees because

of her special **pheromones**.

Queen cage
A special cage in which queens are shipped and/or introduced to a colony, usually with five or six young workers called attendants, and a candy plug.

Queen cell
A special elongated cell resembling a peanut shell in which the queen is reared; usually over 2.5cm (1in) in length, it hangs vertically from the comb.

Queen cup
A cup-shaped cell hanging vertically from the comb, but containing no egg.

Queen excluder
A device which permits workers to pass, but excludes the larger queens and drones.

Queenright
A colony that contains a laying queen.
Radial extractor
See **Extractor**.

Requeen
To introduce a new queen to a queenless hive.

Rendering
The process of melting combs and **cappings** and removing refuse from the wax.

Robbing
The act of bees stealing honey/nectar from other colonies; also applied to bees cleaning out wet **supers** or **cappings** left uncovered by beekeepers.

Ropy characteristic
A diagnostic test for **American foulbrood** in which the decayed larvae form an elastic rope when drawn out with a toothpick or matchstick.

Royal jelly
A highly nutritious, milky white glandular secretion of young (nurse) bees; used to feed the queen and young larvae.

Sacbrood
A **brood disease** of bees caused by a filterable virus which interferes with the moulting process; the dead larva resembles a bag of fluid.

Scout bees
Worker bees searching for a new source of pollen, nectar, propolis, water, or a new home for a swarm of bees.

Sealed brood
See **Capped brood.**

Self-spacing frames
Frames constructed so that they are a **bee space** apart when pushed together in a hive body.

Settling tank
A large-capacity container used to settle extracted honey.

Skep
A beehive without movable frames.

Smoker
A metal container with attached bellows which burns organic fuels to generate smoke.

Solar wax melter or extractor
A glass-covered insulated box used to melt wax from combs and **cappings** using the heat of the sun.

Species of bees
The four most common species of *Apis* are *A. mellifera*, *A. cerana*, *A. dorsata* and *A. florae*. Most of our garden bees are races (subspecies) of *A. mellifera* and other newly discovered races are currently under investigation.

Spermatheca
A small sac in which a queen bee stores spermatozoa.

Split
To divide a colony for the purpose of increasing the number of hives.

Sting
An organ belonging exclusively to female insects developed from egg-laying mechanisms, used to defend the colony; modified into a piercing shaft through which venom is injected.

Sugar syrup
Feed for bees, containing sucrose or table (cane) sugar and hot water in various ratios.

Super
A receptacle in which bees store honey; usually placed over or above the **brood nest**; so-called brood supers contain **brood**.

Supersedure
Rearing a new queen to replace the mother queen in the same hive; shortly after the daughter queen begins to lay eggs, the mother queen disappears.

Swarm
A collection of bees, containing at least one queen, that has split apart from the mother colony to establish a new one; a natural method of propagation of honey bees.

Swarm cell
Queen cells usually found on the bottom of the combs before swarming.

Tangential extractor
See **Extractor**.

Thorax
The central region of an insect to which the wings and legs are attached.

Travelling screen
A framed screen that fits over the top as a hive cover; used when moving bees in hot weather to provide sufficient ventilation to keep bees from suffocating.

Uncapping fork
A fork-like device used to remove wax cappings covering honey, so it can be extracted.

Uncapping knife
A knife used as an alternative to an **uncapping fork**.

Uncapping tank
A container over which frames of honey are uncapped; usually strains out the honey, which is then collected.

Uniting
Combining two or more colonies to form a larger colony.

Varroa
An external mite parasite on honey bees.

Veil
A protective netting that covers the face and neck; allows ventilation, easy movement and good vision.

Virgin queen
An unmated queen bee.

Wasp
A close relative of honey bees, usually in the family Vespidae; they are carnivorous, some species preying on bees.

Wax glands
Eight glands located on abdomen of young worker bees; they secrete beeswax droplets.

Wax moth
Usually refers to the greater wax moth, *Galleria mellonella*, whose larvae bore through and destroy honeycomb as they eat out its impurities.

Western honey bee
See *Apis mellifera*.

Winter cluster
A tight ball of bees within the hive which they form when outside temperature falls below 14°C (57°F) and they need to generate heat.

Worker bees
Infertile female bee whose reproductive organs are only partially developed, responsible for carrying out all the routine of the colony.

Worker comb
Comb measuring about two cells to the centimetre (five to the inch), in which workers are reared and honey and pollen are stored.

Index

Picture Credits

Mic Cady 1; Brian McCallum 2; Brian McCallum 3C; Corbis 3L; Alamy 3R; Brian McCallum 7; Alamy 8–9; Bridgeman Art Library/Bibliotheque Municpale 10; 11; Alamy 12TR; Brian McCallum 12BL; Kim Taylor/The Garden Collection 13BL; Kim Taylor/The Garden Collection 14T, B; Brian McCallum 18; Brian McCallum 19; Brian McCallum 20-21; Jill Mead 22, 23L; Corbis 23R; Alarmy 24; Jill Mead 25; Brian McCullum 27; Corbis 28; Alamy 30; Brian McCallum 27C; Nicola Stocken Tomkins/The Garden Collection 32-33 (Design: Cheryl Waller); E. H Thorne 36, 37, 38; Brian McCallum 42T, B, 51, 55; Brian McCallum 56, 57T, L; Alarmy 58-59; Brian McCallum 66; Alamy 67B; Liz Eddison/The Garden Collection 70-71 (Design: Patrick McCann); Neil Sutherland/The Garden Collection 72; Derek Harris/The Garden Collection 73; Torie Chugg/The Garden Collection 74; Corbis 75; Liz Eddison/The Garden Collection 76-77 (Design: Lucy Hunter); Alarmy 80, 81R, 83; Jill Mead 84; Alamy 85; Brian McCallum 89, 91; Corbis 94; Bridgeman Art Library/Giraudon 95T, B; Corbis 96-97; Jonathan Buckley/The Garden Collection 100 (Design: David & Mavid Seeney); Emjay Smith / Shutterstock 102; Jonathan Buckley/The Garden Collection 904-105 (Design: Helen Yemm, Ketley's); Brian McCallum 107B; Alamy 107T; Brian McCallum 108, 109, Brian McCallum 111T; Alamy 111, 112, 117, 118; Getty Images 119; Punchstock 121; Alamy 123, 124L, R, 125L, R, 126L; iStockphoto 126R; Alamy 127; Brian McCallum 129; Alamy 130; Brian McCallum 131; The Natural History Museum / Alamy 133TR, Paul Airs / Alamy 133BL; Brian McCallum 134; Brian McCalllum 136TL, B, 137; Ann Johansson/Corbis 138, John Walker/ZUMA Press/Corbis 139.

Useful addresses and links

NATIONAL BODIES

Africa
The African Beekeeping Resource Centre
www.apiconsult.com

Australia
Australian Honey Bee Industry Council
www.honeybee.org.au

Canada
Canadian Honey Council
www.honeycouncil.ca
Canadian Association of Professional Apiculturists
www.capabees.ca

France
Union Nationale de L'Apiculture Francaise
www.unaf.net

Ireland
The Federation of Irish Beekeepers' Associations
www.irishbeekeeping.ie

Italy
Istituto Nazionale di Apicoltura
www.inapicoltura.org

New Zealand
National Beekeepers Association of New Zealand
www.nba.org.nz

Norway
Norges Birøkterlag
www.norges-birokterlag.no

South Africa
Southerns Beekeeping Association
http://www.beekeepers.co.za

Sweden
Sveriges Biodlares Riksförbund
www.biodlarna.se

UK
British Bee Keepers Association (BBKA)
BBKA HQ National Beekeeping Centre, National Agricultural Centre, Stoneleigh Park, Warwickshire CV8 2LG
Tel: 02476 696679 Fax: 02476 690682.
www.bbka.org.uk
International Bee Research Association (IBRA)
16 North Road, Cardiff CF10 3DY
www.ibra.org.uk
National Bee Unit
Central Science Laboratory, Sand Hutton, York YO41 1LZ
Tel: 01904 462510
www.beebase.csl.gov.uk/index.cfm

USA
American Beekeeping Federation,
P.O. Box 1337, Jesup, GA 31598-1038
Tel: (912) 427-4233 Fax: (912) 427-8447
www.abfnet.org

MAGAZINES

New Zealand
New Zealand Beekeeper Magazine
www.nba.org.nz/Sections-article7-p1.htm

UK
Journal of Apicultural Research
http://www.ibra.org.uk/jar.html
BBKA News
http://www.bbka.org.uk/articles/bbka-news.php
Bee Craft
www.bee-craft.com

USA
Bee Culture
www.beeculture.com
American Bee Journal
www.dadant.com/journal

EQUIPMENT AND SUPPLIES

Australia
Bindaree Bee Supplies
www.bindaree.com.au
HoneyBee Australis
www.honeybee.com.au

Canada
Benson Bee Supplies
www.bensonbee.com
FWJones & Son Ltd
www.fwjones.com
The Bee Works
www.beeworks.com

USA
Betterbee
www.betterbee.com
Arnold Honeybee Services
www.arnoldhoneybeeservices.com

Denmark
Swienty A/S - Beekeeping Equipment
www.swienty.com

South Africa
Southerns Beekeeping Association
www.beekeepers.co.za

New Zealand
Ecroyd Beekeeping Supplies Ltd
www.ecroyd.com
Ceracell Beekeeping Supplies Ltd
www.ceracellbees.co.nz

UK
E.H.Thorne (Beehives) Ltd
www.thorne.co.uk
National Bee Supplies
www.beekeeping.co.uk
Maisemore Apiaries
www.bees-online.co.uk
Park Beekeeping Supplies
www.parkbeekeeping.com
Stamfordham Ltd
www.stamfordham.biz
Kembles Bee Supplies
www.kemble-bees.com

CHARITIES WORKING WITH DEVELOPING COUNTRIES
Bees for Development
www.beesfordevelopment.org
Bees Abroad
www.beesabroad.org.uk

Other useful links
Bees for Kids
www.bees4kids.org.uk
Apiservices
www.beekeeping.com

Acknowledgments

The authors would like to thank Chris Deaves at the Twickenham and Thames Valley Beekeepers' Association for answering Brian's countless questions; John Chapple, chair of the London Beekeepers' Association for helping Alison with her first colony; John Hamer at Blackhorse Apiaries for enabling us to expand our apiary and knowledge; Jill Mead for her beautiful photographs of urban bees; our parents for all their never-ending support and encouragement, including hosting some hives; friends for their fascination in our new-found hobby; and last but not least Mic Cady for chance meetings on trains and having faith in us to deliver this book.

The publishers would like to thank E. H. Thorne for their help with the hive illustrations, Ted Benton for his patience and contacts in the latter stages of the project and Lizzie Harper for doing the illustrations at such short notice.

Printed in the USA
CPSIA information can be obtained
at www.ICGtesting.com
JSHW072004200524
63503JS00013B/508

9 781446 303559